INTERNATIONAL CENTRE FOR MECHANICAL SCIENCES

COURSES AND LECTURES - No. 317

RELIABILITY PROBLEMS: GENERAL PRINCIPLES AND APPLICATIONS IN MECHANICS OF SOLIDS AND STRUCTURES

EDITED BY

F. CASCIATI
UNIVERSITY OF PAVIA

J. B. ROBERTS
UNIVERSITY OF SUSSEX

SPRINGER-VERLAG WIEN GMBH

Le spese di stampa di questo volume sono in parte coperte da
contributi del Consiglio Nazionale delle Ricerche.

This volume contains 80 illustrations.

Originally published by Springer Verlag Wien-New York in 1991

In order to make this volume available as economically and as rapidly as possible the authors' typescripts have been reproduced in their original forms. This method unfortunately has its typographical limitations but it is hoped that they in no way distract the reader.

ISBN 978-3-211-82319-4 **ISBN 978-3-7091-2616-5 (eBook)**
DOI 10.1007/978-3-7091-2616-5

PREFACE

In many fields of engineering it is necessary to formulate and implement procedures for the assessment of system reliability. This involves an estimation of the probability that a vector of design variables stays within some prescribed safe domain. In some situation it is sufficient to employ a static analysis, taking into account the statistical variability of the system parameters. However, often the dynamic response of an uncertain, time-varying non-linear system to random disturbances must be considered.

Significant advances in reliability theory, during the sixties and seventies have recently been utilized to construct a general design methodology, suitable for incorporating into decision making processes. This is currently finding application in such diverse fields as offshore technology, aerospace design and disaster prevention in civil and mechanical engineering.

Although a basic theoretical framework has been established the inherent techniques require constant development and improvement, to meet the demands imposed by new advanced engineerings projects. The subject is, therefore, still in an active state of development.

This book presents, to researchers and engineers working on problems concerned with the mechanics of solids and structures, the current state of the development and application of reliability methods. The topics covered reflect the need to integrate, within the overall methodology, statistical methods for dealing with systems which have uncertain parameters and random excitation with the development of suitable safety index and design codes. The basic principles of reliability theory, together with current standard methodology, are reviewed. An introduction to new developments is also provided.

F. Casciati

J. B. Roberts

CONTENTS

Chapter 1

RANDOM VIBRATION AND FIRST PASSAGE FAILURE

J. R. Roberts
University of Sussex, Sussex, UK

Abstract

In the first part of this Chapter, a variety of techniques for predicting nonlinear dynamic system response to random excitation are discussed. These include methods based on modelling the response as a continuous Markov process, leading to diffusion equations, statistical linearization, the method of equivalent nonlinear equations and closure methods. Special attention is paid to the stochastic averaging method, which is a combination of an averaging technique and Markov process modelling. It is shown that the stochastic averaging method is particularly useful for estimating the "first-passage" probability that the system response stays within a safe domain, within a specified period of time. Results obtained by this method are presented for oscillators with both linear and nonlinear damping and restoring terms. An alternative technique for solving the first-passage problem, based on the computation of level crossing statistics, is also described: this is especially useful in more general situations, where the stochastic averaging is inapplicable.

1.1 INTRODUCTION

Methods of predicting the response of mechanical and structural systems to random excitation are of importance in many engineering fields: examples include the motion of offshore structures and ships to wave and wind excitation, the response of civil engineering structures, such as buildings, bridges, etc., to earthquakes, the behaviour of vehicles travelling on rough ground and the characteristics of aerospace vehicles when responding to atmospheric turbulence and jet noise excitation.

In many cases one is primarily interested in predicting the reliability of systems responding to random excitation - i.e., predicting the probability that the system will not fail, in some defined sense, within a specified period of time. This almost invariably means that one is concerned with large amplitude response, and hence with motions which involve significant nonlinear effects. In nearly all application areas, the nonlinearities which are inevitably present become of primary importance when the amplitude of the response is large.

Despite extensive research over the last few decades, there is currently no generally applicable theoretical technique for predicting the probabilistic response of complex dynamic systems to random excitation, capable of yielding results with a reasonable degree of computational effort. However, for simple systems some very useful approximate techniques have been developed, which can, in some cases, be used to provide quantitative reliability estimates.

In this Chapter various existing techniques for predicting the response of nonlinear systems to random excitation are briefly described. It is shown that one of these, the method of stochastic averaging, is particularly useful for dealing with nonlinear oscillators, with light damping, driven by wide-band random excitation. Using this technique it is possible to obtain simple explicit expressions for the probability distribution of the response. This information can be directly applied in estimating system reliability.

This discussion is followed by a detailed description of the so-called "first-passage problem" - i.e., the problem of predicting the probability that the response stays within a safe domain (in phase space) within a specified period of time. It is demonstrated that, for nonlinear oscillators, specific analytical solutions can be obtained in some cases, and in other cases it is possible to formulate a simple numerical scheme, based on a random walk analogue. Results obtained by this method are shown to compare well with corresponding digital simulation estimates of first-passage statistics.

The Chapter concludes with a brief description of an alternative approach to estimating first-passage statistics, based on a computation of terms in the so-called inclusion - exclusion series. This approach is useful in situations when the basic assumptions inherent in the stochastic averaging method are inapplicable.

1.2.1 Markov methods

A general form of the equations of motion of an n degree of freedom nonlinear system is as follows [1]

$$\underset{\sim}{M}\,\ddot{\underset{\sim}{q}} + \underset{\sim}{C}\,\dot{\underset{\sim}{q}} + \underset{\sim}{K}\,\underset{\sim}{q} + \underset{\sim}{g}(\underset{\sim}{q},\dot{\underset{\sim}{q}}) = \underset{\sim}{Q}(t) \qquad (1.1)$$

Here $\underset{\sim}{q}$ is an n-vector of generalised displacements and $\underset{\sim}{Q}$ is an n-vector of generalised forces. $\underset{\sim}{M}$, $\underset{\sim}{C}$ and $\underset{\sim}{K}$ are the usual mass, damping and stiffness matrices, respectively and $\underset{\sim}{g}$ contains all the nonlinearities in the system. For simplicity it will be assumed here that the mean of $\underset{\sim}{Q}$ is zero and that $\underset{\sim}{g}$ is odd-symmetric, such that the mean of $\underset{\sim}{q}$ is also zero.

It is convenient to rewrite equation (1.1) in state space form. Introducing the state variable vector $\underset{\sim}{z} = [\underset{\sim}{q}^T, \dot{\underset{\sim}{q}}^T]^T$ one obtains

$$\dot{\underset{\sim}{z}} = \underset{\sim}{F}(\underset{\sim}{z}) + \underset{\sim}{f}(t) \qquad (1.2)$$

where

$$\underset{\sim}{F} = \begin{bmatrix} \dot{\underset{\sim}{q}} \\ -\underset{\sim}{M}^{-1}\underset{\sim}{C}\dot{\underset{\sim}{q}} - \underset{\sim}{M}^{-1}\underset{\sim}{K}\underset{\sim}{q} - \underset{\sim}{M}^{-1}\underset{\sim}{g} \end{bmatrix}, \qquad \underset{\sim}{f} = \begin{bmatrix} 0 \\ \underset{\sim}{M}^{-1}\underset{\sim}{Q} \end{bmatrix} \qquad (1.3)$$

If the elements of f are broad-band in character they can, in many cases, be satisfactorily approximated in terms of stationary white noises. Thus,

$$E\{\underset{\sim}{f}(t)\underset{\sim}{f}^T(t+\tau)\} = \underset{\sim}{D}\,\delta(\tau) \qquad (1.4)$$

where E{ } denotes the expectation operator, δ( ) is the Dirac delta function and $\underset{\sim}{D}$ is a real, symmetric, non-negative matrix. Equation (1.2) can now be written as

$$\dot{\underset{\sim}{z}} = \underset{\sim}{F}(\underset{\sim}{z}) + \underset{\sim}{B}\underset{\sim}{\xi} \qquad (1.5)$$

where ξ is a 2n-vector of independent, unit white noises and $\underset{\sim}{B}\underset{\sim}{B}^T = \underset{\sim}{D}$. It is noted that, since $\dot{z}$ exists almost nowhere [2], this equation needs to be interpreted carefully; a suitable interpretation is to treat it as an Itô equation, written in the form

$$d\underset{\sim}{z} = \underset{\sim}{F}(\underset{\sim}{z})dt + \underset{\sim}{B}d\underset{\sim}{W} \qquad (1.6)$$

where $\underset{\sim}{\xi}(t) = d\underset{\sim}{W}/dt$ and $\underset{\sim}{W}$ is 2n-vector of unit Wiener processes [2].

It follows from equations (1.5) (or (1.6)) that z is a 2n-dimensional Markov process, with a transition density function

[3], $p(\underline{z}|\underline{z}_0;t)$, governed by the following diffusion equation, known as the Fokker-Planck-Kolmogorov (FPK) equation:

$$\frac{\partial p}{\partial t} = Lp \tag{1.7}$$

where

$$L = -\sum_{i=1}^{2n} \frac{\partial}{\partial z_i}\left[F_i(z_i)\,\cdot\right] + \frac{1}{2}\sum_{i,j=1}^{2n}\sum^{2n} D_{ij}\frac{\partial^2}{\partial z_i \partial z_j} \tag{1.8}$$

and

$$\underline{D} = [D_{ij}] = \underline{B}\,\underline{B}^T \tag{1.9}$$

If the system is stable, and time - invariant, a "stationary" solution to the FPK equation usually exists: thus, as time elapses, $p(\underline{z}|\underline{z}_0;t)$ becomes independent of the initial condition and asymptotes towards a stationary density function, $w(\underline{z})$. - i.e.

$$w(\underline{z}) = \lim_{t\to\infty} p(\underline{z}|\underline{z}_0;t) \tag{1.10}$$

$w(z)$ may be obtained as the solution of equation (1.7) with $\partial p|\partial t = 0$ - i.e.,

$$Lw = 0 \tag{1.11}$$

A general, closed-form solution to equation (1.7) has yet to be found, for an arbitary value of n, although series solutions can be found in terms of eigenfunctions and eigenvalues [4]. For $m = 2n = 1$, the system is of first-order and analytical expressions for $p(z|z_0;t)$ have been found in a few special cases [4,5]. However a general expression for $w(z)$ can easily be found, for $m = 1$. The result is

$$w(z) = \frac{c}{D}\exp\left[\frac{2}{D}\int_0^z F(\xi)d\xi\right] \tag{1.12}$$

where c is a normalization constant, chosen so that

$$\int_{-\infty}^{\infty} w(z)dz = 1 \tag{1.13}$$

For second order systems (n=1) one has a single degree of freedom system, or oscillator. A general form of the equation of motion of such systems is

$$\ddot{x} + h(x,\dot{x}) = B\xi(t) \qquad (1.14)$$

Here $z = [z_1, z_2]^T = [x, \dot{x}]^T$ and its transition density is governed by equation (1.7), where here

$$L = -z_2 \frac{\partial p}{\partial z_1} + \frac{\partial}{\partial z_2}(hp) + \frac{B^2}{2}\frac{\partial^2 p}{\partial z_2^2} \qquad (1.15)$$

A complete, exact solution to this equation has been found only for the linear case.

If attention is restricted to the stationary density function, $w(z)$ then a well known result exists for the case where h is of the form

$$h = z_2 H(E) + g(z_1) \qquad (1.16)$$

and E is the total energy of the oscillator - i.e.,

$$E = \frac{z_2^2}{2} + \int_0^{z_1} g(\xi)d\xi \qquad (1.17)$$

The solution is [6]

$$w(z) = c \exp\left\{-\gamma \int_0^E H(\xi)d\xi\right\} \qquad (1.18)$$

where

$$\gamma = 2/B^2 \qquad (1.19)$$

and c is a normalization constant.

For higher order systems very few exact solutions are available. It is possible to generalise equation (1.18) to apply to a certain very specific class of multi-degree of freedom systems, but the applicability of this result is very limited [7].

In view of the paucity of exact solutions, there have been numerous attempts to develop approximate analytical methods, and numerical methods combined with analytical methods. These include iterative methods [8], series expansion methods [4], the use of random walk analogies [9-11], finite element methods [12-13] and the application of path-integral techniques [14].

An important class of approximate methods goes under the name of stochastic averaging [15]; this will be discussed in some detail later in this Chapter, in section 3.

1.2.2 The statistical linearization method

Because linear systems are so much easier to solve than nonlinear ones, a natural approach to attacking nonlinear problems is to replace equations (1.2) by a linear set, of the form

$$\dot{\underset{\sim}{z}} = \underset{\sim}{L}\underset{\sim}{z} + \underset{\sim}{f}(t) \qquad (1.20)$$

where L is a 2n x 2n matrix of constants, and to minimise the equation deficiency

$$\underset{\sim}{\epsilon} = \underset{\sim}{F}(\underset{\sim}{z}) - \underset{\sim}{L}\underset{\sim}{z} \, . \qquad (1.21)$$

If ϵ is minimised in a least square sense - i.e. $E\{\underset{\sim}{\epsilon}^T\underset{\sim}{\epsilon}\}$ is minimised, then one finds that

$$\underset{\sim}{L} = E\{\underset{\sim}{F}(\underset{\sim}{z})\underset{\sim}{z}^T\} \underset{\sim}{V}^{-1} \qquad (1.22)$$

where

$$\underset{\sim}{V} = E\{\underset{\sim}{z}\underset{\sim}{z}^T\} \qquad (1.23)$$

is the covariance matrix for z.

In the case where $f(t)$ is modelled as a white noise vector, such that equation (1.4) holds, then the following Lyanpunov equation for $\underset{\sim}{V}$ can be derived from the equivalent linear system [1]:

$$\underset{\sim}{L}\underset{\sim}{V}^T + \underset{\sim}{V}\,\underset{\sim}{L}^T + \underset{\sim}{D} = \underset{\sim}{0} \qquad (1.24)$$

If the expectation in equation (1.22) is evaluated on the basis that z is Gaussian then, on combining equations (1.22) and (1.24) one obtains a set of nonlinear algebraic equations for the elements of V. These can be solved numerically and, in certain special cases, analytically. The evaluation of L may be simplified under the Gaussian assumption, in terms of expectations of derivatives of $\underset{\sim}{F}$ [1].

The above analysis procedure can be generalised fairly easily to deal with cases where the means are non-zero, and/or the excitation and responses processes are nonstationary [1].

1.2.3 Equivalent nonlinear equation method

A wide class of nonlinear oscillators with white noise excitation are governed by an equation of motion of the form of equation (1.14), where

$$h = b(z_1,z_2) + g(z_1) \qquad (1.25)$$

Unfortunately, in nearly every case of practical interest, $b(z_1,z_2)$ can not be expressed as $z_2H(E)$, as indicated by equation (1.16).

To overcome this difficulty one can replace (1.25) by the equivalent nonlinearity of equation (1.16). minimising the equation deficiency

$$\epsilon = b(z_1,z_2) - z_2H(E) \qquad (1.26)$$

in a least-square sense [16]. When $g(z_1)$ is linear this is clearly a generalization of the statistical linearization procedure, where an equivalent linear system (here H(E) = a constant) is used as the replacement system.

If the damping is light the energy dissipated per "cycle", due to damping is, on average, a small fraction of the average energy in a cycle of oscillation. In this case one can treat the total energy, E, as approximately constant, over one cycle of oscillation. More precisely, over an interval of time corresponding the period, T(E), of free, undamped oscillations, H(E) can be assumed to be approximately constant. The period of free oscillation is given by

$$T(E) = 4\int_0^b \frac{dy}{\{2[E-V(y)]\}^{1/2}} \qquad (1.27)$$

where

$$V(y) = \int_0^y g(\xi)d\xi \qquad (1.28)$$

is the potential energy function and b is such that

$$V(b) = E \qquad (1.29)$$

The error integral

$$I = \int_0^{T(E)} \epsilon^2 dt \qquad (1.30)$$

can be minimised with respect to H(E), where the latter is treated as constant. This yields [17]

$$H(E) = B(E)/C(E) \tag{1.31}$$

where

$$B(E) = \frac{4}{T(E)} \int_0^b b(y, [2(E-V(y)]^{\frac{1}{2}})dy \tag{1.32}$$

and

$$C(E) = \frac{4}{T(E)} \int_0^b [2(E-V(y)]^{\frac{1}{2}}dy \tag{1.33}$$

A combination of equation (1.31) with equation (1.18) now gives

$$w(\underline{z}) = c \exp\left[-\gamma \int_0^E \frac{B(\xi)}{C(\xi)} d\xi\right] \tag{1.34}$$

for the stationary density function of z. This expression will become increasingly accurate as the magnitude of the damping is progressively reduced.

Equation (1.31) can also be derived from energy considerations. Suppose that the energy dissipation per cycle due to the terms $b(z_1,z_2)$ in equation (1.25), and $z_2H(E)$ in equation (1.16) are equated. Thus

$$\int_0^{T(E)} b(z_1,z_2)z_2dt = \int_0^{T(E)} [z_2H(E)]z_2dt \tag{1.35}$$

If H(E) is treated as constant, it is clear that equation (1.35) leads to an equation for H(E) which is identical to equation (1.31).

1.2.4 Closure methods

Equations for the moments of the response can be derived fairly readily from the equations of motion. From these moments, or related quantities known as the cumulants and quasi-moments [1,18] it is possible to derive estimates of the probability distribution of the response, using analytical expansions of the Gaussian distribution.

Thus, if $\underline{z}$ is governed by equation (1.2), where $\underline{f}(t)$ is a white noise vector, with a covariance function matrix given by equation (1.4), then, from Markov process theory one has the following important result [19], for any scalar function $g(\underline{z})$:

$$E\{\dot{g}(\underline{z})\} = E\{\underline{h}^T\underline{F}\} + \tfrac{1}{2}\, tr\{E\{\underline{B}\underline{H}\}\} \qquad (1.36)$$

where

$$\underline{h} = \nabla g = \left[\frac{\partial g}{\partial z_1}, \frac{\partial g}{\partial z_2}, \ldots, \frac{\partial g}{\partial z_N} \right]^T \qquad (1.37)$$

is the gradient vector of g, $\underline{H}$ is the Jacobian matrix of second partial derivatives of g and $tr\{\ \}$ denotes the trace of a matrix. With appropriate choices of g, a variety of moment equations can be generated.

A well known difficulty with this approach is that the moment equations are generally governed by an infinite hierarchy of coupled equations. To obtain a solution it is necessary to introduce a "closure approximation". The simplest level of closure is to assume that the response is Gaussian. It can be shown that this approach leads to results which are identical to those found by the statistical linearization method [1]. Improvements in accuracy can be obtained by resorting to a higher, non-Gaussian level of closure.

1.3 THE STOCHASTIC AVERAGING METHOD

In this section a powerful approximate technique for tackling random vibration problems is introduced, which is particularly suitable when the damping in the system is light. This condition frequently occurs in practice. The method was originally proposed by Stratonovich [18] as a means of obtaining results for nonlinear, self-excited oscillations in the presence of noise. It has subsequently been examined mathematically by various workers [20-22] with a view to establishing a rigorous basis. The technique may be viewed as an extension to the stochastic case of the well-known Bogoliubov and Mitropolsky method for approximately solving deterministic nonlinear vibration problems [23].

As a means of providing a simple introduction to the method, with a minimum of mathematical complexity, the case of an oscillator with nonlinear damping and linear stiffness will be considered first. It will be shown that, if the damping is sufficiently light, and the excitation is broad-band, then the amplitude process, a, of such an oscillator can be modelled as a one-dimensional Markov process. The solution of the FPK equation yields the distribution of a, and, more importantly, the joint distribution of displacement and velocity of the response. From this, various statistics of the response, relevant to reliability, such as level crossing rates, can be computed fairly easily.

Following this discussion, extensions of the method to deal with oscillators with nonlinear stiffness and with multi-degree of freedom systems, will be indicated.

The application of stochastic averaging to mechanical and structural random vibration problems has been reviewed fairly recently by Roberts and Spanos [15].

1.3.1 Oscillators with nonlinear damping and stationary excitation

Consider the following equation of motion of a randomly excited nonlinear oscillator:

$$\ddot{x} + \epsilon^2 h(x,\dot{x}) + \omega_0^2 x = f(t) \tag{1.38}$$

It will be assumed that f(t) is a stationary, broad-band process, with zero mean, which possesses a power spectrum, $S_f(\omega)$. Here $S_f(\omega)$ is defined by

$$S_f(\omega) = \frac{1}{2\pi} \int_{-\infty}^{\infty} w_f(\tau)\cos\omega\tau d\tau \tag{1.39}$$

where $w_f(\tau)$ is the correlation function for f(t).

Attention will now be restricted to the situation where the damping is light - i.e., the scaling parameter ϵ^2 is small. Moreover, it will be assumed that the standard deviation of the stationary response, σ, is such that $\sigma = 0(\epsilon^{-1})$. This is certainly true in the linear case, as can easily be shown by standard linear theory, and appears to be generally true in the nonlinear case, although a rigorous proof seems to be difficult. For analysis purposes it is desirable to scale the excitation so that the level of the response, as measured by σ, is of order ϵ^0. This implies that the excitation spectrum should be of order ϵ^2. This scaling can be made explicit by introducing the process f(t) = $\epsilon z(t)$. The equation of motion now becomes

$$\ddot{x} + \epsilon^2 h(x,\dot{x}) + \omega_0^2 x = \epsilon z(t) \tag{1.40}$$

It is important to appreciate that this step does not imply

restrictions on the "strength" of the excitation process. When ϵ^2 is small the level of the excitation will normally be weak compared with the maximum level of the response: in other words, for light damping, the response will grow until its level is large, in some sense, compared with the level of the excitation. The above scaling simply makes this feature appear explicitly.

1.3.1.1 Transformation of variables

Consider the total energy of the oscillator, E, defined by equation (1.17). Here

$$E = \frac{\dot{x}^2}{2} + \frac{\omega_0^2 x^2}{2} \qquad (1.41)$$

If equation (1.40) is multiplied throughout by $\dot{x}$, equation (1.41) is differentiated throughout with respect to time, and the results are combined, one obtains

$$\dot{E} = P_{in}(t) - P_{dis}(t) \qquad (1.42)$$

where

$$P_{in}(t) = \epsilon\dot{x}z(t) \quad , \quad P_{dis}(t) = \epsilon^2\dot{x}h(x,\dot{x}) \qquad (1.43)$$

Equation (1.42) is simply a power balance equation. In words, it states that the rate of change of the total energy of the oscillator, with respect to time, is equal to the power input due to the random excitation, P_{in}, minus the power dissipation due to the damping mechanism, P_{dis}. Since ϵ is here taken to be small, it follows from equation (1.42) that $\dot{E}$ will also be small - i.e. sample functions of E will be varying slowly with time (at least in a macroscopic sense).

The principal objective of the stochastic averaging method is to average both P_{in} and P_{dis} over time, using the fact that, for small ϵ, E is approximately constant over a period of time corresponding to the natural period of oscillation $T = 2\pi/\omega_0$. In the present example it is convenient to work with an amplitude process a(t), rather than E(t), where a is defined by $E = V(a)$ and $V(.)$ is the potential energy function of the oscillator, here given by $V(x) = \omega_0^2 x^2/2$. Hence $E = \omega_0^2 a^2/2$. a has the same dimensions as x and, when ϵ is small, will "follow" the peaks of the response process, x. Clearly, if E is slowly varying then so is a: hence the latter process can be averaged using the concept previously described in connection with E.

Associated with a is a phase process, ϕ; both a and ϕ may be related to x through the following "Van der Pol transformation":

$$x(t)=a(t)\cos(\omega_0 t+\phi) \quad \dot{x}(t)= -a(t)\omega_0\sin(\omega_0 t+\phi) \quad (1.44)$$

When ϵ is small, both a and ϕ are slowly varying. This can be seen by rewriting the original equation of motion in terms of a and ϕ (see equation (1.40)). This gives

$$\begin{Bmatrix}\dot{a}\\ a\dot{\phi}\end{Bmatrix} = \frac{\epsilon^2}{\omega_0} h(a\cos\Phi, -a\omega_0\sin\Phi)\begin{Bmatrix}\sin\Phi\\ \cos\Phi\end{Bmatrix} - \frac{\epsilon z(t)}{\omega_0}\begin{Bmatrix}\sin\Phi\\ \cos\Phi\end{Bmatrix} \quad (1.45)$$

where $\Phi = \omega_0 t + \phi$.

The right-hand sides of equations (1.45) depend not only on a, ϕ and z but also explicitly on time, through the "oscillatory" terms $\sin\Phi$ and $\cos\Phi$. These terms produce small, relatively rapid fluctuations superimposed on relatively large, but slowly varying fluctuations in a and Φ.

The basic idea of the stochastic averaging method is to simplify by eliminating the fluctuational terms, through a suitable time averaging.

1.3.1.2 Averaging the dissipation term

Consider the special case where the excitation is absent ($z(t) = 0$). The right hand sides of the reduced equations can be averaged by assuming that a and ϕ remain approximately constant, over one cycle. If

$$\begin{Bmatrix}F(a)\\ G(a)\end{Bmatrix} = -\frac{1}{2\pi}\int_0^{2\pi} h(a\cos\Phi, -a\omega_0\sin\Phi)\begin{Bmatrix}\sin\Phi\\ \cos\Phi\end{Bmatrix} d\Phi \quad (1.46)$$

then the averaged equations, may be written simply as

$$\dot{a} = -\frac{\epsilon^2}{\omega_0}F(a), \qquad \dot{\phi} = -\frac{\epsilon^2}{a\omega_0}G(a) \quad (1.47)$$

It can be seen that, in the simplified equations, the equation for a is uncoupled from that of ϕ. Thus the first order equation for a can be solved independently: this, as will be seen, is a basic feature of the averaging method and still applies when the excitation is present. In many cases the damping is an odd function of velocity, such that G(a) is zero; then $\dot{\phi} = 0$, implying that ϕ is a constant.

It is noted that, since a is proportional to ϵ^2, the error involved in treating a and Φ as constants, over a period T, is of order ϵ^2. Thus equations (1.47) are correct to order ϵ^2, with an error of order ϵ^4.

Returning to the more general case where excitation is present the partially simplified forms for equations (1.45) are as follows:

$$\begin{Bmatrix} \dot{a} \\ a\dot{\Phi} \end{Bmatrix} = -\frac{\epsilon^2}{\omega_0}\begin{Bmatrix} F(a) \\ G(a) \end{Bmatrix} - \frac{\epsilon z(t)}{\omega_0}\begin{Bmatrix} \sin\Phi \\ \cos\Phi \end{Bmatrix} \tag{1.48}$$

1.3.1.3 Averaging the excitation term

So far the averaging method corresponds exactly with the deterministic method of Bogoliubov and Mitropolsky [23]. The next step, however, which involves applying averaging to the last terms in equation (1.48), is rather more difficult. The correct procedure for averaging those excitation terms is due to Stratonovich [18]. The basic difficulty in averaging the terms

$$y_1(t) = \frac{1}{\omega_0}z(t)\sin\Phi, \quad y_2(t) = \frac{1}{\omega_0}z(t)\cos\Phi \tag{1.49}$$

lies in the fact that the phase process, Φ, is correlated with z. Thus the means of y_1 and y_2 are generally non-zero, even though the processes z and Φ have zero mean.

Suppose that the excitation process z has a correlation time scale, τ_{cor}, which is so small that one can define a time interval Δt, such that the following two conditions are satisfied, simultaneously:

(i) $\Delta t \gg \tau_{cor}$

(ii) a and Φ do not change appreciably from t to t + Δt.

It is noted that these conditions are always met if ϵ is sufficiently small, for then the correlation time scale associated with the response will always become large, compared with that of the excitation. Put another way, the bandwidth of the response will become small, compared with that of the excitation. From a practical viewpoint, however, it is desirable that the excitation should have a large band-width (so that τ_{cor} is absolutely small), otherwise ϵ may have to be extremely small (and maybe unrealistically so) in order that the above requirements be met.

If the above conditions hold then, after averaging over one

cycle, of period T. the mean of $y_1(t)$ is found to be. approximately, [24]

$$E\{y_1(t)\} \equiv \bar{y}_1 = \frac{-\epsilon\pi S_z(\omega_0)}{2a\omega_0^2} \quad (1.50)$$

where $S_z(\omega)$ is the spectrum of z(t). Moreover, the correlation function of the zero mean process $y^*_1 = y_1 - \bar{y}_1$ is given approximately by [24]

$$R_{y_1}(\tau) = D\,\delta(\tau) \quad (1.51)$$

where the "strength" of the white noise process is given by (after time averaging).

$$D = \pi S_z(\omega_0) \quad (1.52)$$

Thus the process $y_1(t)$ can be approximated as a white noise process, with a strength D given by equation (1.52) and a non-zero mean, given by equation (1.50). Hence one can express $y_1(t)$ as

$$y_1(t) = -\frac{\epsilon\pi S_z(\omega_0)}{2a\omega_0} - [\pi S_z(\omega_0)]^{1/2}\xi_1(t) \quad (1.53)$$

where $\xi_1(t)$ is a zero-mean white noise process, with unit strength. Combining equations (1.48), (1.49) and (1.53) one obtains (noting that $S_f(\omega) = \epsilon^2 S_z(\omega)$)

$$\dot{a} = -\frac{\epsilon^2}{\omega_0}F(a) + \frac{\pi S_f(\omega_0)}{2a\omega_0^2} + \frac{[\pi S_f(\omega_0)]^{1/2}}{\omega_0}\xi_1(t) \quad (1.54)$$

This is the final, "simplified equation" for a(t).

A similar analysis can be carried out for the phase process, $\phi(t)$, to show that

$$a\dot{\phi} = -\frac{\epsilon^2}{\omega_0}G(a) + \frac{[\pi S_f(\omega_0)]^{1/2}}{\omega_0}\xi_2(t) \quad (1.55)$$

where ξ_2 is another white noise process, with zero mean and unit strength, independent of ε_1. It is observed that the mean of the process $y_2(t)$, defined by equation (1.49) is zero. Equation (1.55) is the "simplified equation" for Φ.

It can be shown [18] that the approximations involved in averaging the stochastic terms are entirely consistent with the approximations inherent in averaging the dissipation term. In fact the averaging approximation is, overall, correct to order ϵ^2.

Since a and Φ are governed by a first-order equations, with white noise excitation, it follows that the joint process $[a,\Phi]$ is (at least approximately) a joint Markov process, Strictly, $[a,\Phi]$ converges to a Markov process as ϵ^2 tends to zero. Thus one can regard the simplified equations for a and Φ as asymptotically exact, as $\epsilon\to 0$. The mathematical basis for this result is contained in the Stratonovich-Khasminskii limit theorem [18,20].

1.3.1.4 The FPK equations

With $[a,\Phi]$ approximated as a Markov process, the transition density $p(a,\Phi|a_0,\Phi_0;t)$ can be introduced. Applying equations (1.7) and (1.8) to this case it is found that the transition density function is governed by the following FPK equation:

$$\frac{\partial p}{\partial t} = \frac{\partial}{\partial a}\left[\left\{\frac{\epsilon^2 F(a)}{\omega_0} - \frac{\pi S_f(\omega_0)}{2a\omega_0^2}\right\}p\right] + \frac{\epsilon^2 G(a)}{a\omega_0}\frac{\partial p}{\partial a}$$

$$+ \frac{\pi S_f(\omega_0)}{2a\omega_0^2}\left[\frac{\partial^2 p}{\partial a^2} + \frac{1}{a^2}\frac{\partial^2 p}{\partial \Phi^2}\right] \tag{1.56}$$

An inspection of the differential equations for a and Φ (see equations (1.54) and (1.55)) shows that the amplitude process, a, is uncoupled from the phase process, Φ. It follows that a is a one-dimensional Markov process. The transition density function for a, $p(a|a_0;t)$, is governed by the following FPK equation.

$$\frac{\partial p}{\partial t} = \frac{\partial}{\partial a}\left[\left\{\frac{\epsilon^2 F(a)}{\omega_0} - \frac{\pi S_f(\omega_0)}{2a\omega_0^2}\right\}p\right] + \frac{\pi S_f(\omega_0)}{2\omega_0^2}\frac{\partial^2 p}{\partial a^2} \tag{1.57}$$

1.3.1.5 Stationary solutions

Since the excitation is assumed here to be stationary, the response will approach stationarity as time elapses - i.e.

equations (1.10) and (1.11) apply.

A comparison of the resulting equations reveals that the phase angle ϕ is uniformly distributed between 0 and 2π. Hence

$$w(a,\phi) = \frac{1}{2\pi}\, w(a) \qquad (1.58)$$

where the factor $1/2\pi$ arises from the normalisation condition for $w(a)$ and $w(a,\phi)$.

Through a transformation from a,ϕ variables to the original $x,\dot{x}$ variables, an expression for the stationary density function of x and $\dot{x}$, $w(x,\dot{x})$ can be deduced from equation (1.58). Thus

$$w(x,\dot{x}) = \frac{1}{2\pi\omega_0 a}\, w(a) \qquad (a^2 = x^2 + \dot{x}^2/\omega_0^2) \qquad (1.59)$$

where, from the FPK equation for a, applying the general solution for a first order system given by (1.12),

$$w(a) = ca\, \exp\left\{ -\frac{2\epsilon^2\omega_0}{\pi S_f(\omega_0)} \int_0^a F(\xi)d\xi \right\} \qquad (1.60)$$

where c is a normalisation constant.

In the linear case ($\epsilon^2 h(x,\dot{x}) = 2\varsigma\omega_0\dot{x}$) equations (1.59) and (1.60) lead to the following results:

$$w(a) = \frac{a}{\sigma^2}\, \exp\left\{ -\frac{a^2}{2\sigma^2} \right\} \qquad (1.61)$$

and

$$w(x,\dot{x}) = \frac{1}{2\pi\omega_0\sigma^2}\, \exp\left\{ -\frac{1}{2\sigma^2}\left[x^2 + \frac{\dot{x}^2}{\omega_0^2} \right] \right\} \qquad (1.62)$$

where

$$\sigma^2 = \frac{\pi S_f(\omega_0)}{2\varsigma\omega_0^3} \qquad (1.63)$$

An appropriate integration involving $w(x,\dot{x})$, as given by equation (1.62), shows that σ, as given by equation (1.63) is the stationary standard deviation of x. In the special case of white noise excitation, where $S_f(\omega) = S_0$, a constant, equations (1.61) to (1.63) agree with the well-known exact results for this case [1]. According to equation (1.61), a has a Rayleigh distribution whereas, from equation (1.63), the joint distribution of x and $\dot{x}$ is Gaussian.

It is also of interest to compare the above approximate solutions with known exact solutions for nonlinear oscillators excited by white noise. One such exact solution exists for the case where the nonlinearity is of the form (see equations (1.16) to (1.19), with H(E) a constant)

$$\epsilon^2 h(x,\dot{x}) = 2\zeta\omega_0\dot{x} + \epsilon^2 g(x) \qquad (1.64)$$

In this case one finds that

$$\epsilon^2 F(a) = a\zeta\omega^2{}_0 \qquad (1.65)$$

just as in the case of linear stiffness. Thus, according to the stochastic averaging theory, the small nonlinear stiffness term does not contribute to the stationary response distribution. However, as one would expect, the exact solution for white noise excitation (equation (1.18)) shows that the nonlinearity in stiffness can markedly affect the distribution of the response. This apparent anomaly is easily resolved when it is recalled that the stochastic averaging solution is an approximation which is only correct to order ϵ^2. If the exact solution is expanded in powers of ϵ^2, and terms of order higher than ϵ^2 are discarded, then one finds that the nonlinear stiffness does indeed vanish. Thus the stochastic averaging approximation is consistent with the exact solution.

In the case where the stiffness is linear and the damping has the specific form $\epsilon^2\rho(a)\dot{x}$, where $\rho(a)$ is some arbitary function of a, then the stochastic averaging theory gives a result for $w(x,\dot{x})$ in complete agreement with the exact solution (equation (1.18)).

1.3.2 Oscillators with nonlinear stiffness

We now consider an oscillator with light, nonlinear damping and a nonlinear restoring force. The equation of motion is

$$\ddot{x} + \epsilon^2 h(x,\dot{x}) + g(x) = f(t) \qquad (1.66)$$

For this system the total energy E, given by equation (1.17) is again governed by equations (1.42) and (1.43). On suitably generalising the averaging theory given in section 2 it is possible to show [24] that, if ϵ is small and $f(t)$ is modelled as a white noise, E is approximately a one-dimensional Markov process, governed by the first order equation

$$\dot{E} = H(E) - J(E)\xi(t) \tag{1.67}$$

where

$$H(E) = - B(E) + D[1-C'(E)/2]/2 \tag{1.68}$$

$$J(E) = [DC(E)]^{1/2} \tag{1.69}$$

and $B(E)$ and $C(E)$ are given by equations (1.32) and (1.33), respectively. (here $b = \epsilon^2 h$).

Using standard methods [2] the FPK equation for $p(E|E_0;t)$, the transition density function for E, is given by

$$\frac{\partial p}{\partial t} = \frac{\partial}{\partial E}\left\{\left[\epsilon^2 B(E) - \frac{D}{2}\right]p\right\} + \frac{D}{2}\frac{\partial^2}{\partial E^2}[C(E)p] \tag{1.70}$$

where D is the "strength" of $f(t)$ (see equation (1.4)).

1.3.2.1 Stationary solution

The stationary density for E may be found by solving equation (1.11). This gives

$$w(E) = cT(E)\exp\left\{-\frac{2Q(E)}{D}\right\} \tag{1.71}$$

where

$$Q(E) = \int_0^E \frac{B(\xi)}{C(\xi)}\,d\xi \tag{1.72}$$

and c is a normalization constant. In the case of linear stiffness this is equivalent to a result found earlier, using standard stochastic averaging (see equation (1.60)).

The relationship between $w(a)$ and $w(x,\dot{x})$, given by equation (1.59) can also be generalised. By a variety of physical arguments [18] it can be shown that

$$w(x,\dot{x}) = \frac{w(E)}{T(E)} \tag{1.73}$$

Hence

$$w(x,\dot{x}) = k \exp\left[-\frac{2}{D}Q\left[\frac{\dot{x}^2}{2} + V(x)\right]\right] \qquad (1.74)$$

On comparing this result with that obtained earlier, by the method of equivalent nonlinear equations (see equation (1.34)), and noting that $D = B^2$, it can be seen that the results are, in fact, identical.

1.4. THE FIRST PASSAGE PROBLEM

In practical applications involving the random vibration of a mechanical or structural system it is often required to estimate the probability that the system's response will stay within safe, prescribed limits, within a specified interval of time. The determination of such a probability is usually called the "first-passage problem", and has been extensively studied, particularly during the last two decades [25].

1.4.1 First-passage statistics

In practical applications, there is usually a "safe region" of operation, the outer limits of which are defined by a suitable "barrier". The first-passage problem may be stated thus: To find the probability, $P(t)$, that a response process, $x(t)$, of a system crosses some critical barrier (i.e. exits from the safe domain) at least once in some interval of time 0-t.

It is noted that $P(t)$ will depend on the initial condition of the system, at $t = 0$. Associated with $P(t)$ is the "reliability function" $Q(t) = 1-P(t)$.$Q(t)$ is the probability that $x(t)$ stays within the safe domain, in the interval 0-t.

The first-passage density function $p(t)$, is defined such that $p(t)dt$ is the probability that $x(t)$ first exceeds the barrier between times t and $t+dt$. It is easy to show that $p = dP/dt$.$p(t)$ is simply the density function of the random time, T, to first-passage failure. The moments of T may be expressed as

$$M_n = E(T^n) = \int_0^\infty t^n p(t)dt \qquad (1.75)$$

Of these moments the first, M_1, which is simply the mean time to failure, is by far the most important. For stationary excitation it will be shown later that

$$Q(t) \to \exp(-\alpha t) \quad \text{as } t \to \infty \tag{1.76}$$

where α is called the "limiting decay rate". When M_1 is very large, equation (1.76) is a good approximation for nearly all values of t and it follows that

$$\alpha \to 1/M_1 \quad \text{as } M_1 \to \infty \tag{1.77}$$

In this connection it is noted that for a Gaussian process the distribution of barrier crossings is asymptotically Poisson distributed as the critical level (b) becomes very large [26], where ν is the average number of barrier crossings (from within the safe region, to outside the region) per unit of time. This implies that

$$\alpha \to \nu \quad \text{as } b \to \infty \tag{1.78}$$

Two types of barrier are of primary importance - single-sided and double-sided barriers. The first of these is such that any value of x(t) less than a fixed level, b say, is safe. Thus first-passage failure occurs when x(t) first exceeds b. The second type is such that, for safe operation $|x(t)| < b$.

A third type of barrier - the "circular barrier" - is of considerable theoretical interest. Here the safe domain is within a circle in the phase-plane, $(x, \dot{x}/\omega_0)$, of radius b; thus

$$b^2 = x^2 + \dot{x}^2/\omega_0^2 \tag{1.79}$$

1.4.2 The "exact" approach

We now consider an oscillator with the following equation of motion:

$$\ddot{x} + h(x,\dot{x}) = f(t) \tag{1.80}$$

Here x(t) is the displacement response, $h(x,\dot{x})$ is a general function of displacement and velocity, and f(t) will be assumed to be a zero-mean, stationary white noise process.

1.4.2.1 Diffusion equations

As discussed earlier the joint process $z = [x(t), \dot{x}(t)]$ is a two-dimensional continuous Markov process, with a transition probability density function, governed by an FPK equation (see equations (1.7) and (1.8). In the context of first-passage problems it is more convenient to consider the conditional transition density function $q(z|z_0;t)$, where $q(\underset{\sim}{z}|\underset{\sim}{z}_0;t)$ is the probability that $z < z(t) < z + dz$ at time t, without departing from the safe domain. Both conditional and unconditional transition density functions are governed by the same FPK equation. Thus

$$\frac{\partial q}{\partial t} = \frac{\partial}{\partial y}[h(x,y)q] - y\frac{\partial q}{\partial x} + \frac{D}{2}\frac{\partial^2 q}{\partial y^2} \qquad (1.81)$$

where $y = \dot{x}$ and D is the "strength" of the white noise process (see equation (1.4)).

The FPK equation can be considered as a continuity equation for the "flow" of "probability mass". This is evident if equation (1.81) is re-written as

$$\frac{\partial q}{\partial t} + \frac{\partial}{\partial x}(F_x) + \frac{\partial}{\partial y}(F_y) = 0 \qquad (1.82)$$

where

$$F_x = yq \quad , \quad F_y = -h(x,y)q - \frac{D}{2}\frac{\partial q}{\partial y} \qquad (1.83)$$

F_x is the flow of probability mass in the x-direction and F_y is the corresponding flow in the y-direction. To solve equation (1.81) one requires a suitable initial condition. Normally a deterministic condition is appropriate - i.e. $\underset{\sim}{z}(0) = \underset{\sim}{z}_0$. Thus

$$\lim_{t \to 0} q(\underset{\sim}{z}|\underset{\sim}{z}_0;t) = \delta(\underset{\sim}{z}-\underset{\sim}{z}_0) \qquad (1.84)$$

A "stationary start" condition is also sometimes appropriate; here it is assumed that there is an ensemble of initial values, $\underset{\sim}{z}_0$, distributed according to the stationary response distribution.

Solving equation (1.81) enables one to determine how the probability mass "diffuses" with time in the phase plane, from some initial condition. The diffusion process is, alternatively, governed by an integral equation (the Chapman-Kolmogorov-Smoluchowski (CKS) equation); this may be written as follows:

$$q(\underset{\sim}{z}|\underset{\sim}{z}_0;t) = \int_R q(\underset{\sim}{z}|\underset{\sim}{z}';t-t')q(\underset{\sim}{z}'|\underset{\sim}{z}_0;t')d\underset{\sim}{z}' \qquad (1.85)$$

where the integration range, R is the safe domain in the phase plane. Equations (1.84) and (1.85) are equivalent mathematical representations of the same underlying diffusion law for the process $\underset{\sim}{z}$.

Associated with the FPK equation is the adjoint form, usually referred to as the backward Kolmogorov equation. This is of the form

$$\frac{\partial q}{\partial t} = \mathcal{L} q \tag{1.86}$$

where $\mathcal{L}$, the "backward operator", is here given by

$$\mathcal{L} = -h(x_0, y_0) \frac{\partial}{\partial y_0} + y_0 \frac{\partial}{\partial x_0} + \frac{D}{2} \frac{\partial^2}{\partial y_0^2} \tag{1.87}$$

This equation can be used to derive a differential equation for the reliability function $Q(t|\underset{\sim}{z}_0)$ - this is the probability that first-passage will not occur in the interval 0-t, for trajectories in the phase-plane starting at $\underset{\sim}{z}_0$. Clearly,

$$Q(t|\underset{\sim}{z}_0) = \int_R q(\underset{\sim}{z}|\underset{\sim}{z}_0;t)d\underset{\sim}{z} \tag{1.88}$$

Thus, by integrating equation (1.84) with respect to $\underset{\sim}{z}$, over R, one obtains

$$\frac{\partial Q}{\partial t} = \mathcal{L} Q \tag{1.89}$$

$P(t|\underset{\sim}{z}_0)$ is governed by the same differential equation as $Q(t|\underset{\sim}{z}_0)$.

From equation (1.89) it is fairly easy to derive the following set of differential equations (usually called the generalised Pontriagin-Vitt (GPV) equations)

$$M_n = -nM_{n-1} \qquad (n = 0,1,2...) \tag{1.90}$$

Since $M_0 = 1$ equations (1.90) can be solved successively, to yield M_1, M_2, etc. Of principal interest is the Pontriagin-Vitt (PV) equation for the mean time to failure, M_1, obtained by setting $n = 1$ in equation (1.90).

1.4.2.2 Boundary conditions

For a single-sided barrier, it is observed that the flow of probability mass in the x direction, F_x (see equations (1.83)) is such that flow from the safe, to the unsafe region, at $x = b$, occurs only if $y > 0$. For $x = b$ and $y < 0$, F_x is negative, indicating a "return flow", from the unsafe to the safe region. To ensure that the return flow does not occur it is sufficient to set $q = 0$ for $x = b$, $y < 0$. Similar arguments can be used to specify boundary conditions for double-sided and circular barriers. It can be shown that these conditions lead to well-posed problems [27].

An alternative approach is to consider solving the differential equation governing Q, as given by equation (1.89). For the single-sided barrier, for example, plausible boundary conditions for Q are as follows:

$$\left.\begin{array}{lll} Q(0|x_0,y_0) & = 1 & \underset{\sim}{z}_0 \in R \\ Q(t|b,y_0) & = 0 & y_0 > 0 \\ Q(t|x_0,y_0) & \rightarrow 0 & |y_0| \rightarrow \infty \end{array}\right\} \quad (1.91)$$

The first condition expresses the fact that, within a zero time interval, the probability of not exiting from the safe domain is unity (for starts within the safe domain). The second condition corresponds to the fact that if the diffusion process starts on the line $x = b$, $y > 0$, the oscillator response will immediately move out of the safe domain, and thus the probability of not exceeding the barrier is zero. The third condition implies that, if the velocity is infinite, first-passage failure is bound to occur, within any given time interval. It can be shown that these boundary conditions lead to a well-posed problem for the determination of Q. Similar boundary conditions can be postulated for double-sided and circular barriers [25].

1.4.2.3 <u>Exact analytical solutions</u>

For two-dimensional diffusion processes exact solutions have so far been obtained for only very special, reduced cases where one or more terms in the equation of motion are neglected.

For the mean time to failure of a randomly accelerated free particle, in the case of a double-sided barrier, a complete analytical solution has been obtained in terms of hypergeometric functions [28].

In view of the complexity of this solution it is, perhaps, not surprising that exact analytical solutions are not available for the more general case of non-zero $h(x,y)$, even for just the mean time to failure, M_1.

In a quite different approach, a result due to Dynkin [29] may be used to obtain specific exact results, in some circumstances. If $H(x,y)$ is some function of x and y then Dynkin's formula may be written as

$$E\{H(x,y)_t\} - E\{H(x,y)_{t_0}\}$$

$$= E\left\{ \int_{t_0}^{t} \mathcal{L} H[x(s),y(s)]\, ds \right\} \qquad (1.92)$$

where $\mathcal{L}$ is the backward operator and t is any random time. If t is the first-passage time, and $\mathcal{L}[h(x,y)]$ is just a constant, C, then,

$$E\left\{ \int_{t_0}^{t} \mathcal{L} H[x(s),y(s)]\, ds \right\} = C[E\{t\} - E\{t_0\}] \qquad (1.93)$$

and it follows that equation (1.92) gives an expression for the mean time to first-passage failure, $M_1 = E\{t\}$.

As an illustration, consider the case of an undamped oscillator. Let $h(x,y) = E$, where E is the total energy, defined by equation (1.17). Using equation (1.87) one finds that

$$\mathcal{L}[h(x,y)] = \frac{D}{2} \qquad (1.94)$$

and hence equation (1.92) will give an expression for M_1. For a deterministic initial start condition, i.e., $\underline{z}(0)$ and $E(0)$ are known at $t_0 = 0$, then the result is

$$E\{E(T) - E(0)\} = \frac{D}{2} M_1 \qquad (1.95)$$

If T is the time to reach a constant energy level, E, then $E\{E(T)\} = E$ and equation (1.95) gives

$$M_1 = \frac{2}{D} [E - E(0)] \qquad (1.96)$$

It is interesting to note that, in this simple and exact result, the stiffness function does not appear explicitly.

1.4.3 Approximate analytical solutions

If the domain of safe operation is a closed region then it is possible to solve the PV equation for M_1 by a Galerkin technique [30].

Consider the general case of an nth order system, with state variables $\underset{\sim}{x} = [x_1, x_2, \ldots, x_n]^T$, driven by white noise. Here the PV equation for M_1 is once again given by equation (1.90), with n = 1, where relates to the n state variables. An appropriate boundary condition is $M_1 = 0$ on the boundary of the safe domain.

The Galerkin method is to seek an approximate solution in the form of a truncated series.

$$M_1(\underset{\sim}{x}) = \sum_{k=1}^{m} M^{(k)} \phi_k(\underset{\sim}{x}) \qquad (1.97)$$

where ϕ_k is some system of functions which are complete within the safe domain and satisfy the boundary condition. A set of linear algebraic equations for the constants $M^{(k)}$ can then be generated. Results have been obtained, by this means, for linear oscillators with a circular barrier and for a system relating to the response of a thin curved panel to random excitation [30].

1.4.4 Numerical solutions

The first numerical solutions of the first passage problem, for the case of a simple linear oscillator, were obtained by solving the CKS equation (see equation (1.85)) numerically [31]. This amounts to diffusing probability mass over a discretised phase-plane, using discrete time steps; the unconditional transition density function, $p(\underset{\sim}{z}|\underset{\sim}{z}_0;t)$, was used to redistribute the probability mass at each time step, and probability mass which diffused outside the safe domain was counted as "lost".

For nonlinear oscillators this approach can not be used, as the transition density function $p(\underset{\sim}{z}|\underset{\sim}{z}_0;t$ is generally not known. However, a discrete random walk analogue can be constructed, which can be viewed as a finite difference approximation of the governing FPK equation for $q(\underset{\sim}{z}|\underset{\sim}{z}_0;t)$. Results have been obtained by this technique for various types of nonlinear oscillators. [29]

A natural alternative to finite difference schemes is the use of finite elements. Numerical solutions to both equations (1.89) and (1.90), have been obtained by using the Petrov-Galerkin finite element method [32]. A feature of the finite element numerical method, as with other numerical methods, is that the computational effort increases quite sharply as the barrier level increases. In fact the number of finite elements required increases as the barrier level squared, and the cost of computation increases as the fourth power of the barrier level.

1.4.5 Stochastic averaging approximations

For an oscillator with light nonlinear damping. and linear stiffness, it was shown earlier that an appropriate equation of motion is given by equation (1.40) and that the amplitude process, a, where (see equation (1.44))

$$a = \left[x^2 + y^2/\omega_0^2 \right]^{\frac{1}{2}} \tag{1.98}$$

converges to a Markov process, as $\epsilon \to 0$. The limiting stochastic equation for a is given by equation (1.54).

The conditional transition density function for $a(t)$, $q(a|a_0;t)$, is governed by the following FPK equation (which is the same equation that governs the unconditional transition density function, $p(a|a_0;t)$ - see equation (1.57):

$$\frac{\partial q}{\partial t} = \frac{\partial}{\partial a} \left[\left\{ \frac{\epsilon^2 F(a)}{\omega_0} - \frac{\pi S_f(\omega_0)}{2a\omega_0^2} \right\} q \right] + \frac{\pi S_f(\omega_0)}{2\omega_0^2} \frac{\partial^2 q}{\partial a^2} \tag{1.99}$$

The associated backward equation is given by equation (1.86). where here

$$\mathcal{L} = - \left[\frac{\epsilon^2 F(a)}{\omega_0} - \frac{\pi S_f(\omega_0)}{2a_0\omega_0^2} \right] \frac{\partial}{\partial a_0} + \frac{\pi S_f(\omega_0)}{2\omega_0^2} \frac{\partial^2}{\partial a_0^2} \tag{1.100}$$

From this, differential equations for the reliability function Q. and the moments, M_n, of T can be derived. They are of the general form of equations (1.89) and (1.90), where now the one-dimensional operator, $\mathcal{L}$, is given by equation (1.100). In this case. the GPV equations for M_n are ordinary differential equations. and are thus much easier to solve then their two-dimensional counterparts.

In the particular case of linear damping $\epsilon^2 h(x,y) = 2\varsigma\omega_0 y$. Introducing a non-dimensional amplitude, A and a non-dimensional time τ, as follows

$$A = a/\sigma \quad :, \quad \tau = \omega_0 t \tag{1.101}$$

where σ is the standard deviation of the stationary response, $x(t)$, as given by equation (1.63). then, equations (1.54), (1.99) and (1.100) may be written in the following non-dimensional form:

$$\dot{A} = - \varsigma \left[A - \frac{1}{A} \right] + \sqrt{2}\varsigma \, \xi(t) \tag{1.102}$$

$$\frac{1}{\varsigma}\frac{\partial q}{\partial \tau} = -\frac{\partial}{\partial A}\left[\left[A - \frac{1}{A}\right]q\right] + \frac{\partial^2 q}{\partial A^2} \tag{1.103}$$

$$\mathcal{L} = \varsigma\left[-\left[A - \frac{1}{A}\right]\frac{\partial}{\partial A} + \frac{\partial^2}{\partial A^2}\right] \tag{1.104}$$

1.4.5.1 Boundary conditions

Of the three types of barrier considered earlier, it is clear that the circular barrier is the easiest to deal with, when working with the amplitude process, $a(t)$. Thus the probability that the response stays within the circular barrier is simply the probability that $a(t)$ stays below the level b, where $x^2+y^2/\omega_0{}^2 = b^2$. For the FPK equation the appropriate boundary conditon for $q(a|a_0;t)$ is given by

$$q = 0 \qquad \text{for } a = b \tag{1.105}$$

whereas, for the reliability function $Q(t|a_0)$ one must have

$$\left.\begin{aligned} Q(0|a_0) &= 1 \qquad a_0 < b \\ Q(t|b) &= 0 \qquad t < 0 \end{aligned}\right\} \tag{1.106}$$

The boundary conditions for $M_n(a_0)$ are

$$M_n(b) = 0 \tag{1.107}$$

In addition it is necessary to ensure that $Q(t|0)$ and $M_n(0)$ are finite.

In the case of the single-sided barrier, failure will occur when $a(t) = b \sec\theta(t)$ (see Fig. 1.1(a). This rather complicated condition can be simplified by replacing the original barrier by the one shown in Fig. 1.1(b). Here the segments PA, PB of the original barriers are folded back, to lie along the horizontal axis. For light damping, response trajectories will be roughly circular and the probability that a trajectory reaches the barrier in Fig. 1.1(c), without crossing the barrier in Fig. 1.1(a), and vice-versa, is negligible. Thus first-passage times for the barriers shown in Figs. 1.1(a) and (b) should be almost identical if the damping is light.

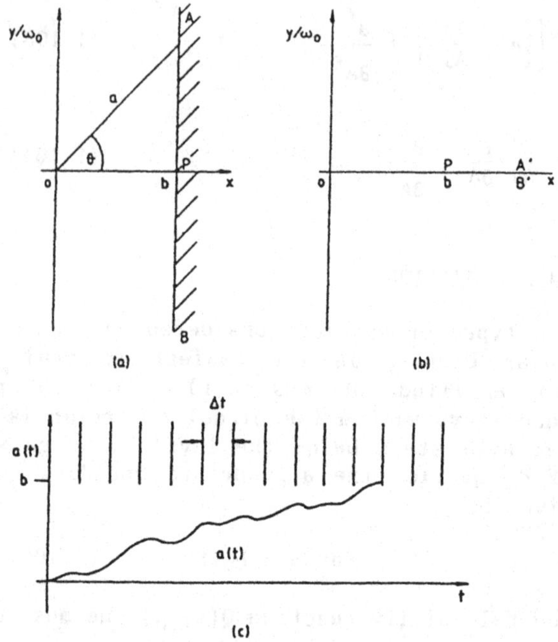

Fig. 1.1 Boundary conditions for a(t)

For the modified single-sided barrier failure occurs only when $\theta(t) = 0$, and the times at which this occurs will have a small dispersion about the equi-spaced times $t_n = 2n\pi/\omega_0$. Thus, a reasonable approximation is to adopt the barrier shown in Fig. 1.1(c), for the a(t) process. Here failure occurs only at the discrete times, t_n. This is equivalent to replacing the continuous time envelope (CE process) by a discrete time envelope (DE process), for the purpose of first-passage analysis. For the DE process one can write.

$$q(a|a_0;t_{n+1}) = \int_0^b p(a|a';\Delta t)q(a'|a_0;t_n)da' \quad (1.108)$$

(compare with equation (1.85) for the two-dimensional case), where $\Delta t = 2\pi/\omega_0$. If $p(a|a';\Delta t)$ is known, equation (1.108) can be marched in time, from some prescribed initial condition, to determine the evolution of $q(a|a_0;t)$. Hence the survival probability

$$Q(t|a_0) = \int_0^b q(a|a_0;t)da \quad (1.109)$$

can be computed, at every time step. It is noted that the vertical barrier lines in Fig. 1.1(c) act as "absorbers" of probability mass, which block further diffusion.

For a double-sided barrier, very similar approximations can be made and one is led to equation (1.108) again, where now $\Delta t = \pi/\omega_0$

- i.e. one half of the time interval appropriate for a single-sided barrier.

From the nature of these approximations for single and double-sided barriers, and the fact that $p(a|a_0;t)$ depends on the product $\epsilon^2 t$, two important conclusions can be drawn regarding the behaviour of the resulting estimates of $Q(t|a_0)$ (and hence $P(t|a_0)$).

(1) $Q(t|a_0)$ will depend only on b, $\mu\epsilon^2$ and $\epsilon^2 t$, where $\mu = 1$ for single-sided barriers and $\mu = 1/2$ for double-sided barriers.

(2) $Q(t|a_0)$ values for the DE process approach the corresponding values of the CE process as $\epsilon^2 \to 0$.

1.4.5.2 Exact analytical solutions

From an analytical viewpoint the simplest case to consider is the circular barrier. Here a separation of variables technique can be used to find a series solution to the first-passage problem, in terms of the eigenfunctions and eigenvalues of the FPK equation for a [24].

In general an exact analytical solution is very difficult to obtain. However, in the case of a linear oscillator a complete analytical solution can be found in terms of hypergeometric functions. This leads to the following expression for $Q(\tau|A_0)$ [25]

$$Q(\tau|A_0) = \sum_{i=1}^{\infty} D_i \exp(-2\varsigma\gamma_i\tau) \tag{1.110}$$

where

$$D_i = \Phi_i\left[\frac{A_0^2}{2}\right] \frac{\int_0^{\eta^2/2} \Phi_i(x)e^{-x}dx}{\int_0^{\eta^2/2} \Phi_i^2(x)e^{-x}dx} \tag{1.111}$$

$$\Phi_i(x) = M(-\gamma_i;1;x) \quad , \quad \eta = b/\sigma \tag{1.112}$$

and γ_i are the roots, or eigenvalues, of the equation

$$\Phi(\eta^2/2) = 0 \tag{1.113}$$

If the mean time to first passage is large then. as in the general case, equation (1.110) reduces to the limiting form of equation (1.76), where here

$$\alpha = 2\varsigma\omega_0\gamma_1 \qquad (1.114)$$

The moments, M_n. of the first passage time T, can be obtained directly from the general solution given by equation (1.110). Of most practical interest is the first moment, $M_1(A_0)$ – the mean time to failure – and the variance of the time to failure

$$\mathrm{var}\{T|A_0\} = M_2(A_0) - M_1^2(A_0) \qquad (1.115)$$

$M_1(A_0)$ can be expressed as

$$M_1(A_0) = \frac{1}{2\varsigma}\left[\bar{E}_i\left[\frac{n^2}{2}\right] - \bar{E}_i\left[\frac{A_0^2}{2}\right] - \ln(n^2/A_0^2)\right] \qquad (1.116)$$

where $\bar{E}i(\ )$ is the exponential integral function [33].   The variance is given by

$$\mathrm{var}\{T|A_0\} = \frac{1}{\varsigma^2}\sum_{i=1}^{\infty}\int_0^{A_0/\sqrt{2}} \frac{(4n-1)\exp(x^2)}{n(2n)!} \ dx \qquad (1.117)$$

In the special case of zero damping, the solution to the PV equation for M_1 agrees with the exact result given earlier (see equation (1.96)).

In the more general case of a nonlinear oscillator with a circular barrier and an equation of motion given by equation (1.80) exact solutions for $Q(t|A_0)$ are not available. However. if M_1 is large, useful asymptotic results for this statistic can be deduced from the appropriate PV equation [25]. Since $\alpha \to 1/M_1$ as $M \to \infty$, the asymptotic behaviour of M_1 also gives the asymptotic behaviour of $Q(t|a_0)$.

If the PV equation for M_1 is written in self-adjoint form it can be integrated directly to yield the expression [34]

$$M_1(a_0) = \frac{2\omega_0^2}{\pi S_f(\omega_0)}\int_{a_0}^{b} \frac{W(a)}{w(a)} \ da \qquad (1.118)$$

where

$$W(a) = \int_0^a w(\xi)d\xi \qquad (1.119)$$

and $w(a)$ is the stationary solution of the FPK equation. For large b the integrals in these equations can be evaluated asymptotically, in specific cases, to yield analytical expressions for M_1.

For problems involving single or double-sided barriers it is necessary to solve equation (1.108). If a separable solution

$$q(a|a_0:t) = \Phi(a)\Psi(t) \qquad (1.120)$$

is assumed then one finds that [35]

$$\Phi(a) = \lambda\int\Phi(a')p(a|a':\Delta t)da' \qquad (1.121)$$

where

$$\lambda = \Psi(t_n)/\Psi(t_{n-1}) \qquad (1.122)$$

Thus the general solution to equation (1.108) may be written as

$$q(a|a_0:t_n) = \sum_{i=1}^{\infty} \Phi_i(a)e^{-\gamma_i t_n} \qquad (1.123)$$

where

$$\gamma_i = \frac{1}{\Delta t}\ln\lambda_i \qquad (1.124)$$

λ_i is the ith eigenvalue of equation (1.121) and $\Phi_i(a)$ is the corresponding eigenfunction.

In general, analytical solutions of the eigenvalue problem, represented by equation (1.121), are not available for linear or nonlinear oscillators. However, in the limiting case where M_1 becomes very large, the first term of the summation of equation (1.123) becomes dominant and hence $Q(t|a_0)$ approaches the asymptotic form of equation (1.76). Moreover, it can be shown that $\alpha \to \nu$ as $n \to \infty$ [36]. As mentioned earlier, this is probably the correct, exact asymptotic limit for the limiting decay rate.

1.4.5.3 Semi-analytic solutions and approximations

One approach to finding solutions in the nonlinear case, for the circular barrier, is to seek approximate solutions of the form [37]

$$Q(\tau|A_0) = \sum_{i=1}^{m} c_i(\tau)\Phi_i\left[\frac{A_0^2}{2}\right] \qquad (1.125)$$

where $\Phi_i(\ )$ are the set of eigenfunctions relating to the linear solution (see equation (1.112), $c_i(\tau)$ are functions of time and m is an integer. The expansion of equation (1.125) is particularly convenient since the eigenfunctions are orthogonal with respect to e^{-x}, and the eigenvalues are known.

By using this Galerkin technique a linear set of differential equations for $c_i(\tau)$ can be determined and solved using standard methods. The resulting estimate of $Q(\tau|A_0)$ will progressively improve in accuracy as the number of terms, m, in the summation is increased.

1.4.5.4 Numerical solutions

For a nonlinear oscillator, with a circular barrier, the moments $M_1(a_0)$ of the time to first-passage failure can be obtained fairly easily, by numerically solving the sequence of ordinary GPV equations. Results for the first moment have been obtained by this means [32] for the case of an oscillator with combined linear and quadratic damping.

For single and double-sided barriers, equation (1.108) is the governing equation. As mentioned earlier, this can be solved numerically by marching in time, with steps Δt, provided that $p(a|a_0;\Delta t)$ is known analytically. An eigenfunction expansion for this transition density function can be written. However, for the class of oscillators under discussion an analytical solution for the eigenfunction problem is known only for the case of linear damping. This solution leads to the expression

$$p(a|a_0;\Delta t) = \frac{a}{\beta} \exp\left\{ -\frac{(a^2 + a_0^2 e^{-2\varsigma\omega_0\Delta t})}{2\beta} \right\} I_0\left[\frac{aa_0 e^{-2\varsigma\omega_0\Delta t}}{\beta}\right] \qquad (1.126)$$

where

$$\beta = \alpha^2[1 - \exp(-2\varsigma\omega_0 t)] \qquad (1.127)$$

and I_0 is the modified Bessel function of zero order [34].

Equation (1.108) has been solved numerically, for the case of a linear oscillator, by performing numerical integrations at each time step, and using equation (1.126) [35]. It was found that $q(a|a_0;t_n)$ becomes proportional to the first eigenfunction, $\Phi_1(a)$, as n becomes large, and that

$$R(n) = \frac{Q(n)}{Q(n+1)} \rightarrow \lambda_1 \tag{1.128}$$

as $n \rightarrow \infty$, where

$$Q(n) = \int_0^b q(a|a_0;t_n)da \tag{1.129}$$

Thus the first eigenvalue (and hence the limiting decay rate) can be estimated by marching in time, until the ratio R(n) reaches its limiting value. For $\eta < 3$ this technique is satisfactory but for higher barrier levels λ_1 is so close to unity that the degree of computational effort required to produce accurate estimates of α is prohibitive. This difficulty can be overcome by expanding $q(a|a_0;t)$ in terms of the eigenfunctions of $p(a|a_0;t)$, which are in the form of Laguerre polynomials. This technique leads to a recurrence relationship for the unknown coefficients in the expansion, which are easily solved [35].

Results have been presented for $P(t|a_0)$ and α, for a range of barrier levels, η and damping factors ζ, and compared with corresponding results obtained by using the circular barrier. Similar results for α were reported earlier [39]. Empirical expressions for these results for α to allow easy extension to other values of η and ζ have been given [40].

For oscillators with nonlinear damping a discrete random walk analogue, $R(t_j)$, for the process a(t) can be constructed. For this purpose the FPK equation governing a can be written in the standard form

$$\frac{\partial p}{\partial t} = -\frac{\partial}{\partial a}\left\{A(a)p\right\} + \frac{B^2}{2}\frac{\partial^2 p}{\partial a^2} \tag{1.130}$$

where

$$A(a) = -\frac{\epsilon^2 F(a)}{\omega_0} + \frac{\pi S_f(\omega_0)}{2a\omega_0^2}, \quad B = \frac{\pi S_f(\omega_0)}{\omega_0^2} \tag{1.131}$$

A and B are related to the "infinitesimal moments of a, defined by

$$\alpha_n\,(a,t,\Delta t) = E\{[a(t) - a(t-\Delta t)]^n\} = E\{\Delta a^n\} \tag{1.132}$$

where Δa is the change in $a(t)$, during an interval of time δt. Using the transition density function $p(a|a_1:t)$, α_n can be expressed as

$$\alpha_n(a,t,\Delta t) = \int_0^\infty d\xi(\xi-a)^n p(\xi|a;\delta t) \tag{1.133}$$

From standard Markov process theory one can show that, if $a(t)$ is a Markov process,

$$A(a) = \lim_{t\to 0} \frac{\alpha_1(a,t,\Delta t)}{\delta t}, \quad B = \lim_{t\to 0} \frac{\alpha_2(a,t,\Delta t)}{\delta t} \tag{1.134}$$

are finite whilst

$$\lim_{t\to 0} \frac{\alpha_n(a,t,\delta t)}{\delta t} = 0 \qquad \text{for } n > 2 \tag{1.135}$$

It is possible to find a variety of discrete random processes which possess the same infinitesimal moments in the limit where the time step approaches zero - i.e. which converge to a continuous diffusion process, described by the FPK equation given by equation (1.130).

Suppose that $R(t_j)$ is a random walk process, such that it can only assume the discrete amplitudes a_k, where $a_k = k\delta a$ ($k = 0,1,\ldots,$). If it is in the state $)a_k, t_j)$ then it will be assumed that it can only move to state (a_{k+1}, t_{j+1}) with probability r_k, or to state (a_{k-1}, t_{j+1}) with probability $q_k = 1-r_{kj}$. This is illustrated in Fig. 1.2(a). A typical sample function of $R(t_j)$ is shown in Fig. 1.2(b). The moments of this process are

$$\left.\begin{aligned} \alpha_1(a_k,t_j,\delta t) &= (r_k-q_k)\delta a \\ \alpha_2(a_k,t_j,\delta t) &= r_k(\Delta a)^2 + q_k(\delta a)^2 = (\Delta a)^2 \end{aligned}\right\} \tag{1.136}$$

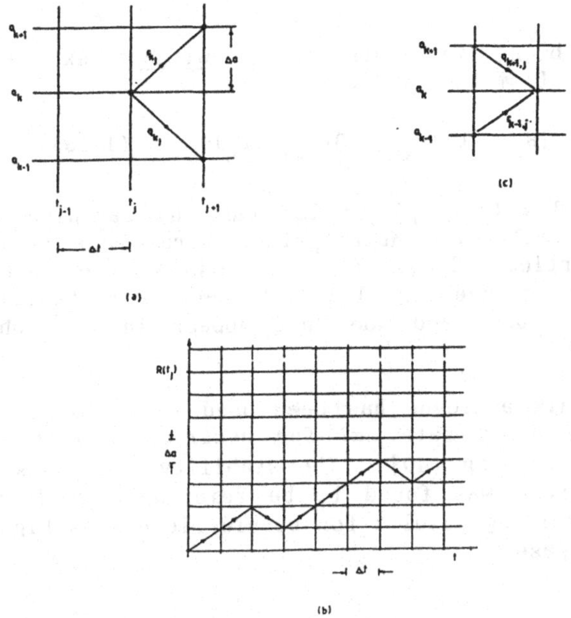

Fig. 1.2 Random walk analogue for a(t)

The limiting random walk process, as $\delta t \to 0$, can be arranged to have the same infinitesimal moments as a(t) by setting

$$(r_{kj}-q_{kj})\delta a = A(a_k)\delta t \ , \ \delta a^2 = B\delta t \qquad (1.137)$$

Since $r_k = 1-q_k$, equation (1.137) can be used to calculate r_k and q_k. Thus,

$$r_k = \frac{1}{2}\left[1 + \frac{A(a_k)}{B}\delta a\right] \ , \ q_k = \frac{1}{2}\left[1 - \frac{A(a_k)}{B}\right]\delta a \quad (1.138)$$

In implementing the random walk scheme one is essentially diffusing probability "mass", over a discrete mesh of space and time. With a deterministic start condition, a(t) = a_k (say), at t = 0, one begins with a unit probability mass, located at (a_k,0), and diffuses this mass forward in time, using the scheme illustrated in Fig. 1.2(a). Special care needs to be taken at a = 0, since the expressions for r_k and q_k become singular here [41]. This difficulty can be overcome by treating a = 0 as a reflecting barrier; thus probability mass is reflected off this barrier, ensuring that there is no diffusion of mass into the region a < 0. One should also be careful in the choice of δa and δt, to ensure that r_k and q_k remain positive, and hence that the random walk process remains physically meaningful. The random walk scheme can easily be shown to be equivalent to a particular finite difference approximation of the FPK equation [40].

The random walk analogue can be used to compute the evolution of the reliability function, Q(t). Thus, if $Q(a_k,t_n)$ is the probability of being in state a_k, t_n, without exceeding the barrier, one has, from the random walk model,

$$Q(a_k,t_{n+1}) = Q(a_{k-1},t_n)r_{k-1} + Q(a_{k+1},t_n)q_{k+1} \quad (1.139)$$

This result enables $Q(a_k,t_n)$ to be found by marching in time steps, δt, (see Fig.2.1(b)) When $t_n = i\Delta t$, where Δt is the interval between the vertical lines of the approximated single or double-sided barrier (see Fig. 1.2.(c)) then the probability mass is absorbed, for $a_k > b$, and does not appear in the subsequent diffusion process.

The random walk analogue has been used to estimate $P(t|a_0)$, and the limiting decay rate, α, for oscillators with combined linear and cubic damping [40]. The stability and robustness of the numerical scheme was found to be remarkably good, with no difficulty in obtaining results for barrier levels as high as η = 6, in the linear case.

1.4.6 Use of the energy envelope

It was pointed out earlier that for oscillators governed by equation (1.66) the energy envelope E(t) could be approximated accurately as a one-dimensional Markov process, as $\epsilon \to 0$. The conditional transition density function $q(E|E_0;t)$ for the Markov model of E(t) is governed by the same FPK equation as the unconditional transition density function (see equation (1.70) and the backward operator, $\mathcal{L}$, for E(t) is

$$\mathcal{L} = -\left[\epsilon^2 B(E_0) - \frac{D}{2}\right]\frac{\partial}{\partial E_0} + \frac{D}{2}C(E_0)\frac{\partial^2}{\partial E_0^2} \quad (1.140)$$

The easiest first-passage problem to solve, when dealing with E(t), is to find the probability, $Q(t|E_0)$ that E(t) stays below a critical level, E, in the interval 0-t, given that $E(0) = E_0$, a known initial value. This is a natural generalisation of the circular barrier problem considered earlier, for the special case where g(x) is linear. If the damping is sufficiently light then $Q(t|E_0)$ will be a good approximation to the probability that x(t) stays below the level b, in the interval 0-t, where E = V(b).

A complete analytical solution for $Q(t|E_0)$ can, in principle, be obtained in terms of an eigenfunction expansion. However, the appropriate eigenfunction problem has been solved, so far, only for case of linear damping and a power-law spring - i.e.

$$g(x) = k|x|^{\nu}\mathrm{sgn}(x) \quad (1.141)$$

where sgn(x) is the sign fuction and k and ν are constants. In this case the functions B(E) and C(E) are both proportional to E so the eigenfunction problem is not dissimilar from that in the linear case. This solution for $Q(t|E_0)$ [42] generalises the result given earlier for the linear case where $\nu = 1$ (see equations (1.110) to (1.113)).

The moments, $M_n(E_0)$, of the time it takes for E(t) to first reach E, starting from E_0, can be determined fairly easily by solving the GPV equations recursively, for n = 1,2,... etc. [42]. In the case of a linear damper and a power-law spring, the moments may be found analytically. In other cases the GPV equations may be solved numerically, without difficulty, since they are ordinary differential equations. Specific results have been obtained for the Duffing oscillators, with linear damping [42].

In the special case of zero damping ($\epsilon^2 = 0$) the PV equation for $M_1(E_0)$ becomes

$$\frac{I}{2}\frac{dM_1}{dE_0} + \frac{D}{2}C(E_0)\frac{d^2M_1}{dE_0^2} = -1 \qquad (1.142)$$

By inspection the solution (subject to the appropriate boundary conditions) is in complete agreement with the exact solution given by equation (1.96).

The diffusion equations for E(t) may be used to find approximate solutions to the single and double-sided barrier problems for the displacement process x(t). This involves a generalisation of the arguments given earlier, for a(t), and leads to the following discrete-time diffusion equation:

$$q(E|E_0:t_{n+1}) = \int_0^h p(E|E':\Delta t)q(E|E_0:t_n)dE' \qquad (1.143)$$

where h = V(b). This is clearly a generalisation of equation (1.108) for a(t) . $q(E|E_0:t_n)dE$ is the probability that E(t) reaches the differential element dE centred at E, at time t_n, without intersecting the vertical barrier lines, spaced Δt apart.

In the case of a nonlinear spring the choice of an appropriate value for Δt in equation (1.143) is not obvious. It has been shown [36,43] that it is best to choose as follows:

$$\Delta t = \mu T(h) \qquad (1.144)$$

where, as before, h = V(b) and $\mu = 1$ for single-sided barriers,

1/2 for double-sided barriers. $T(E)$ is the undamped natural period of the oscillator (see equation (1.27)). With this choice of Δt it can be shown that $\alpha \to \nu$ as $b \to \infty$ [36].

Equation (1.143) can be marched in time, with step length Δt, provided that the unconditional transition density function is known. Available analytical solutions for this function are limited to the class of oscillators with linear damping and a nonlinear spring of power-law form [38]. Estimates of the limiting decay rate, for some oscillators in this class, have been obtained by numerically evaluating the evolution of $q(E|E_0;t)$, according to equation (1.144)). For high barrier levels an expansion technique for $q(E|E_0;t)$ is required, to ensure numerical stability [37].

In cases where $p(E|E_0;t)$ is not known analytically a random walk analogue of the diffusion equation can be used, which is a generalisation of the analogue discussed earlier, for the case where $g(x)$ is linear. Results obtained by this method have been presented for the case of a Duffing oscillator with linear damping [36]. Alternatively, an implicit finite difference approximation can be used to solve the FPK equation governing $q(E|E_0;t)$ [43]. As with the random walk analogue, this enables one to march the diffusion process forward in time, with steps, δt. However, the latter approach has the advantage that δt is no longer necessarily proportional to the square of the amplitude increment. Thus larger steps in time can be used, without sacrifice of accuracy.

1.4.7 Nonstationary problems

The case of purely external excitation will now be returned to. If this excitation is considered to be a modulated white noise

$$f(t) = \alpha(t)\xi(t) \qquad (1.145)$$

where $\alpha(t)$ is a deterministic modulating function and $\xi(t)$ is stationary white noise, of unit strength, then the "zero-start" solutions discussed earlier in this paper relate to the case where $\alpha(t)$ is a step function.

If the approximate energy envelope is adopted then, if $\alpha(t)$ is some arbitrary function, D in the diffusion equation for $E(t)$ can be replaced by $\alpha^2(t)$. Numerical solutions for the diffusion equations can easily be generalised to allow for time-dependency of $\alpha^2(t)$. For example, the discrete-time envelope method can be used to obtain extimates of the probability of first passage failure of a linear oscillator, where

$$\alpha(t) = A[\exp(-c_1 t) - \exp(-c_2 t] \quad ; \quad t > 0 \qquad (1.146)$$

and A, c_1 and c_2 are constants [34].

In cases where the stochastic averaging method is applicable (i.e. g(x) linear, or nearly so) a rather more general model of the nonstationary excitation, f(t), can be adopted, using the concept of the evolutionary power spectrum [15]. Using this spectral representation one simply replaces $S(\omega_o)$ by $S(\omega_o,t)$, in the differential generator $\mathcal{L}$, for example [15].

Whilst numerical solutions are easily modified to deal with modulated excitation, analytical solutions are much more difficult. One semi-analytical approach is to use step-function modulated excitation as the basis for a Galerkin method of solution. Results for a linear oscillator with modulated excitation, and a circular barrier, have been obtained [44].

1.4.8 Level crossing methods

In cases where the response envelope can not be adequately approximated as a Markov process, an alternative approach is possible, using level crossing statistics. In principle, at least, many relevent level crossing statistics can be computed if the response is assumed to be Gaussian.

1.4.8.1 Series solution

Consider the functions

$$Q_n(t) = \underset{0<t_1<t_2<\ldots<t_{n-1}<t}{\int^t \int^t \ldots \int^t} p(t_1,t_2,\ldots,t_{n-1},t)dt_1dt_2\ldots dt_{n-1} \qquad (1.147)$$

where $p(t_1,t_2,\ldots,t_n)$ is the probability that a barrier crossing (from inside to outside) occurs in each of the intervals $dt_1,dt_2\ldots dt_n$. If the last barrier crossing is the kth ($k \geqslant n$) then the preceding n-1 time intervals $dt_1,dt_2,\ldots,dt_{n-1}$ must contain n-1 of the k-1 crossings which occur prior to t. Each distinct way in which these n-1 crossings can occur contributes a term $p_k(t)$ to the total integral $Q_n(t)$, where $p_k(t)dt$ is the probability that x(t) crosses the barrier for the kth time in the interval t,t+dt. The number of distinct ways of choosing a sequence of n-1 crossings from k-1 crossings is just

$${}^{k-1}C_{n-1} = \frac{(k-1)!}{(k-1)!\ (n-1)!} \qquad (1.148)$$

so that

$$Q_n(t) = \sum_{k=n}^{\infty} {}^{k-1}C_{n-1}\, p_k(t) \qquad (1.149)$$

An expression for $p_1(t)$ may be obtained in terms of $Q_n(t)$ by successively eliminating $p_2(t)$, $p_3(t)$, etc. Hence one obtains

$$p_1(t) = \sum_{n=1}^{\infty} (-1)^{n+1} Q_n(t) \qquad (1.150)$$

This is the inclusion-exclusion series of Rice [45].

In practice, the evaluation of $Q_n(t)$ for large n proves difficult: one is forced to use a truncated form of this series to yield an approximation to $P_1(t)$. If m terms are used in the truncated series one has

$$p_1(t) = \sum_{n=1}^{m} (-1)^{n+1} Q_n(t) + (-1)^m \sum_{k=m+1}^{\infty} p_k(t)\, {}^{k-2}C_{m-1} \qquad (1.151)$$

Since the remainder term is positive if m is even and negative if m is odd the truncated series gives an upper bound to $p_1(t)$ if m is odd and a lower bound if m is even.

Using the truncated form of the series expansion, the first-passage probability, P, is given by

$$P(T) = \sum_{n=1}^{\infty} (-1)^{n+1} \int_0^T Q_n(t)\,dt \qquad (1.152)$$

Numerical results for a linear oscillator driven by Gaussian white noise excitation have been obtained for m = 3 [45].

Another serious disadvantage of equation (1.150) is that, for long intervals of time. T, the terms in the series become increasingly large, as n increases. At first sight this restricts its use to small T values. However. this difficulty can be reduced by recasting equation (1.150) into the following sum [46]:

$$P(T) = 1-\exp\left\{\sum_{n=1}^{\infty} \frac{(-1)^n}{n!} I_n\right\} \qquad (1.153)$$

where

$$I_n = \int_0^T\int_0^T \cdots \int_0^T g_n(t_1,t_2,\ldots,t_n)dt_1dt_2\ldots dt_2 \qquad (1.154)$$

The relationship between the g_n and p_n functions is obtained by expanding the right hand side of equation (1.154), and comparing the result with equation (1.150).

The g_n are symmetric functions of their arguments. For all real processes X(t) with a finite "memory" time they have the important property that their value falls to zero when the times $t_1, t_2, \ldots, t_n$ are moved apart - i.e.,

$$g_n\ (t_1,t_2,\ldots,t_n)\to 0$$

provided that at least one difference t_i-t_j of the arguments is increased without limit ($|t_i-t_j|\to\infty$). This does not apply to the p_n functions which, under the same conditions, reduce to products of lower order p_n functions.

1.4.8.2 Limiting distribution

If the response process, x(t), is stationary some simplification can be acheived for the integrals I_n defined by equation (1.155).

For n=1, $g.(t) = \nu$, a constant (the average crossing rate). Clearly

$$I_1 = \nu T \qquad (1.155)$$

ν can be calculated easily using Rice's formula [45]:

$$\nu = \int_0^{\infty} |\mathring{x}|w(a,\mathring{x})d\mathring{x} \qquad (1.156)$$

where $w(x,\mathring{x})$ is the joint density function for x and $\mathring{x}$.

For n = 2, $g_2(t_1,t_2)$ depends only on the difference between t_1 and t_2. Hence $g_2(t_2-t_1) = g_2'(\tau)$, where $\sigma_1 = t_1-t_2$ and

$$I_2 = \int_0^T\int_0^T g_2(t_1,t_2)dt_1dt_2 = 2\int_0^T (T-\sigma_1)g_2'(\sigma_1)d\sigma_1 \quad (1.157)$$

This last integral can be generalised. It can be shown that, for n>2, [46]

$$I_n = \int_0^T (T-\sigma_1)K_n(\sigma_1)d\sigma_1 \quad (1.158)$$

where

$$K_n(\sigma_1) = n(n-1)\int_0^{\sigma_1}\int_0^{\sigma_1}\dots\int_0^{\sigma_1} g_n'(\sigma_1,\sigma_2,\dots,\sigma_{n-1})d\sigma_2 d\sigma_3\dots d\sigma_{n-1} \quad (1.159)$$

Here

$$g_n(t_1,t_2,\dots,t_n) = g_n'(\sigma_1,\sigma_2,\dots,\sigma_{n-1}) \quad (1.160)$$

for a stationary process, where

$$\sigma_i = t_i - t_n \quad (i=1,2,\dots,n-1) \quad (1.161)$$

Further simplifications ensure when T is large. Equation (1.158) can be written in the form

$$I_n = J_n T\left\{1 - \frac{\tau_n}{T}\right\} \quad (1.162)$$

where

$$J_n(T) = \int_0^T K_n(\sigma_1)d\sigma_1 \;,\; \tau_n(T) = \frac{1}{J_n}\int_0^T \sigma_1 K_n(\sigma_1)d_1 \quad (1.163)$$

The integrals J_n and τ_n will approach asymptotic values, J_n^* and τ_n^* respectively, when T becomes large. τ_n^* are measures of the range of σ_1 over which $K_n(\sigma_1)$ are effectively non-zero. τ_n^* can thus be regarded as a sequence of integral time scales.

For high critical levels it has been shown [46] that a maximum finite τ_n^* exists. If T is very much larger than the maximum τ_n^* exists. It follows from the previous equations than an appropriate expression for P(T) is:

$$P(T) = 1-\exp(-\alpha T) \quad (1.164)$$

where

$$\alpha = \gamma - \sum_{n=2}^{\infty} \frac{(-1)^n}{n!} J_n^* \qquad (1.165)$$

Equation (1.164) is in agreement with equation (1.76) for Q(t); thus α here is the limiting decay rate, introduced earlier, in section 1.4.1.

1.4.8.3 Estimation of the limiting decay rate

The computation of J_n^* for n>3 is prohibitively time consuming. Thus, from equation (1.165).

$$\alpha = \gamma[1 - A_2 + A_3 + \text{higher order terms}] \qquad (1.166)$$

where the non-dimensional terms $A_2 = J_2^*/2\nu$, $A_3 = J_3/6\nu$ can be computed, if x(t) is normally distributed [46]. In cases where the higher order terms in equation (1.165) are not negligible, a "closure" approximation, based on A_2 and A_3 is required.

Stratonovitch [46] has proposed a "non-approaching points" closure method based on the first two terms in the series solution. This involves replacing the true g_n functions, for n>2 by approximations based on g_1 and g_2. It leads to the result

$$P(T) = 1-\exp\left\{ \int_0^T \frac{\ln[1-B(T,\tau)]}{B(T,\tau)} g_1(\tau)d\tau \right\} \qquad (1.167)$$

where

$$B(T,\tau) = -\int_0^T \frac{g_2(\tau,u)}{g_1(\tau)} du \qquad (1.168)$$

For a stationary process, x(t), equation (1.168) becomes

$$B(T,\tau) = -\frac{1}{\nu}\int_0^T g_2'(\tau-u)du \qquad (1.169)$$

If T is very large compared with τ_2^*, $B(T,\tau)$ is virtually independent of τ, for all τ values, except those very near to the end points of its range, and has the value

$$B(T,\tau) = -\frac{2}{\nu}\int_0^\infty g_2'(\sigma)d\sigma = -2A_2 \qquad (1.170)$$

It follows that, when T is large, equation (1.167) reduces to

$$P(T) = 1 - \exp\left\{ - \nu T \left[\frac{\ell n[1+2A_2]}{2A_2}\right]\right\} \qquad (1.171)$$

This is the same limiting form given by equation (1.164). Here the limiting decay rate is given by

$$\frac{\alpha}{\nu} = \frac{\ell n[1+2A_2]}{2A_2} \qquad (1.172)$$

On expanding the log term as a power series, it is found that

$$\frac{\alpha}{\nu} = [1-A_2 + \frac{4}{3} A_2^2 - 2A_2^3 + \ldots] \qquad (1.173)$$

An alternative approach is to apply a transformation to equation (1.166). For example, an Euler transformation may be used [46]: this gives

$$\alpha = \nu \; [0.5 + (1-A_2)/4 + (1-2A_2+A_3)/8 + \text{higher order terms}] \qquad (1.174)$$

This transformed series will converge more rapidly than the original series. Similarly, a non-linear transformation due to Shanks [46] may be applied, to give

$$\alpha = \nu \left\{ \frac{A_2 + A_3 - A_2^2}{A_2 + A_3} + \text{higher order terms} \right\} \qquad (1.175)$$

For a geometric series the higher order terms is in equation (1.175) are zero: they are often very small even when the original series is only a rough approximation to a geometric series.

1.4.9 Comparison with simulation

To illustrate the accuracy of some fore-going methods some comparisons with simulation results will be presented, for the case of a linear oscillator excited by white noise. The non-dimensionalised limiting decay rate, α/ν is plotted against non-dimensional barrier height η, for various values of $\mu\varsigma$ (μ defined earlier).

Figs. 1.3(a) and (b) show a comparison between simulation results for a single sided barrier, with $\mu\varsigma$ = 0.04 and 0.005 respectively, compared with theoretical results obtained by the discrete time Markov envelope/(DE) method (full line) and the non-approaching points (NAP) closure method, derived from level crossing statistics (broken line) (see [35] for further details). The DE results are seen to be excellent agreement with simulation estimates. The NAP estimates are less accurate at low values of η. On the other hand the NAP method has the advantage that it has a wider range of applicability.

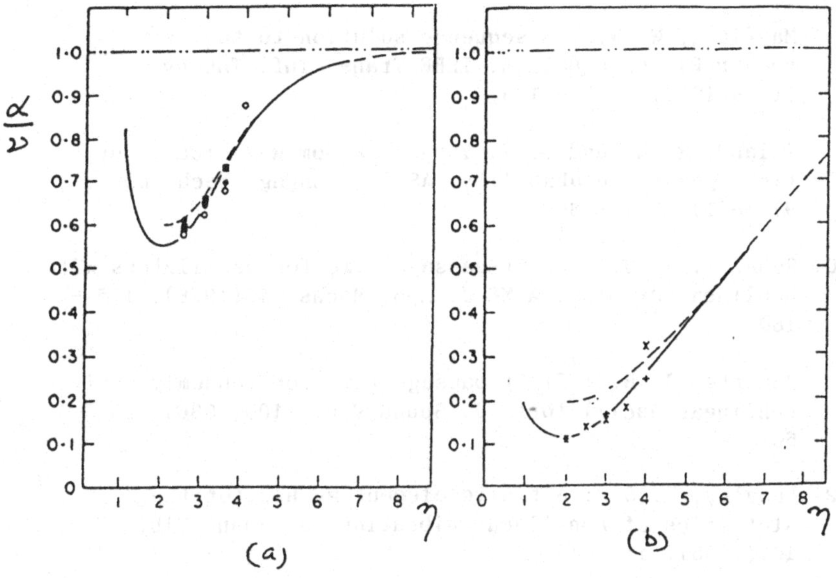

Fig. 1.3 Variation of α/ν with η for a single-sided barrier. (a)

REFERENCES

1. Roberts, J. B. and P. D. Spanos : Random Vibration and Statistical Linearization, J. Wiley, Chichester 1990.

2. Arnold, L. : Stochastic Differential Equations : Theory and Applications, Wiley Interscience, New York 1973.

3. Roberts, J. B. : Response of nonlinear mechanical systems to random excitation : Markov methods, Shock Vib. Digest, 13(1981), 17 - 28.

4. Atkinson, J. D. : Eigenfunction expansions for randomly excited non-linear systems, J. Sound Vib., 30(1973), 153 - 172.

5. Caughey, T. K. : Nonlinear theory of random vibrations, Adv. Appl. Mechs., 11(1971), 209 - 253.

6. Caughey, T. K. : On the response of a class of nonlinear oscillators to stochastic excitation. Proc. Coll. Int. Du Centre Nat. de la Recherche Scient., No 148. Marseille, France (1964), 393 - 402.

7. Dimentberg, M. F. : Statistical Dynamics of Nonlinear and Time-Varying systems, Research Studies Press. Taunton, U.K. (1988).

8. Mayfield, W. W. : A sequence solution to the Fokker-Planck equation. IEEE Trans. Inf. Theory. IT-19(1973), 165 - 175.

9. Toland, R. H. and C. Y. Yang : Random walk model for first-passage probability, ASCE J. Engng. Mech. Div., 97(1971). 791 - 806.

10. Roberts, J. B. : First-passage time for oscillators with nonlinear damping. ASME J. App. Mechs, 45(1978). 175 - 180.

11. Roberts, J. B. : First passage time for randomly excited nonlinear oscillators. J. Sound Vib.. 109(1986). 33 - 50.

12. Langley, R. S. : A finite element method for the statistics of non-linear vibration. J. Sound Vib.. 101(1985). 41 - 49.

13. Bergman, C. A. and Spencer. B. F. : First passage of sliding rigid structures on a frictional foundation. Earthquake Engng. Struct. Dyns. 13(1985), 281 - 291.

14. Wehner. M. F. and W. G. Wolfer : Numerical evaluation of path integral solutions to the Fokker-Planck equations. Phys. Rev., A27(1983), 2663 - 2670.

15. Roberts. J. B. and P. D. Spanos : Stochastic averaging : an approximate method for solving random vibration problems. Int. J. Non-linear Mechs., 21(1986). 111 - 134.

16. Caughey, T. K. : On the response of nonlinear oscillators to stochastic excitation. Prob. Engng. Mechs.. 1 (1986), 2 - 4.

17. Hennig, K. and J. B. Roberts : Averaging methods for randomly excited nonlinear oscillators. in Random Vibration - Status and Recent Developments (Ed. R. H. Lyon) Elsevier, Amsterdam 1986, 143 - 161.

18. Stratonovitch, R. L. : Topics in the Theory of Random Noise [2 vols], Gordon and Breach, New York 1964.

19. Jazwinskii, A. : Stochastic Processes and Filtering Theory, Academic Press, New York (1970).

20. Khasminskii, R. Z. : A limit theorem for the solutions of differential equations with random right-hand sides, Theory Prob. Apps. 11 (1966). 390 -405.

21. Khasminskii, R. Z. : On the averaging principle for stochastic differential Itô equations, Kibernetika, 9 (1968), 260 - 279.

22. Papanicolaou, G. G. and W. Kohler : Asymptotic theory of mixing stochastic ordinary differential equations, Comna Par. App. Maths. 27 (1974), 641 - 668.

23. Bogoliubov, N. and A. Mitropolsky: Asymptotic Methods in the Theory of Non-linear Oscillations, Gordon and Breach, New York 1961.

24. Roberts, J. B. : Averaging Methods in Random Vibration, Dept. Struct. Engng., Technical Univ. Denmark, Report. R 245, 1989.

25. Roberts, J. B. : First-Passage Probabilities for randomly excited systems - diffusion methods, Prob. Engng. Mechs., 1 (1986), 66 - 81.

26. Cramer, H.: On the intersections between the trajectories of a normal stationary stochastic process and a high level, Arkiv fur Matematik, 6(1966), 337-349.

27. Yang, J - N. and M. Shinozuka : First-passage time problem, J. Acoust. Soc. Am., 47 (1970), 393 - 394.

28. Franklin, J.N. and E.R. Roderich: Numerical analysis of an elliptic-parabolic partial differential equation, SIAM J. Num. An., 5(1968), 680-716.

29. Dynkin, E.B.: Markov Processes, Springer-Verlag, New York, 1965.

30. Bolotin, V.V.: Statistical aspects in the theory of structural stability, Proc. Int. Conf. on Dyn. Stability of Structs., Northwestern Univ., Ill., Pergamon Press, (1965) 67-81.

31. Crandall, S.H., K.L. Chandiramini and R.G. Cook: Some first-passage problems in random vibration. ASME J. App. Mechs., 33(1966), 532-538.

32. Bergman, L.A. and B.F. Spencer: Solution of the first passage problem for simple linear and nonlinear oscillators by the finite element method, Dept. Theor. & App. Mechs., Univ. Ill. at Urbana - Champ., Rep. No. 461, 1983.

33. Gradshteyn, I.S. and I.M. Ryzhik: Tables of Integrals, Series and Products, Academic Press, San Diego, 1980.

34. Seshadri, V., B.J. West, and K. Lindenberg: Analytic theory of extrema II. Application to nonlinear oscillators, J. Sound Vib., 68(1980, 553-570.

35. Roberts, J.B.: First passage time for the envelope of a randomly excited linear oscillator, J.Sound Vib., 46(1976), 1-14.

36. Roberts, J.B.: First passage time for oscillators with nonlinear restoring forces, J. Sound Vib., 56(1978), 71-86.

37. Spanos, P.D.: Survival probability of nonlinear oscillators subjected to broad-band random distrubances, Int. J. Nonlinear Mechs., 17(1982), 303-317.

38. Roberts, J.B.: Probability of first passage failure for lightly damped oscillators. Proc. IUTAM Symp. Stochastic Problems in Dynamics (ed: B.L. Clarkson), Pitman 1976.

39. Mark, W.D.: On false-alarm probabilities of filtered noise, Proc. IEEE, 54(1966), 316-317.

40. Lutes, L.D., Y.T. Chen and S. Tzuang: First passage approximation for simple oscillators, ASCE J. Engng. Mechs. Div., 106(1980), 1111-1124.

41. Roberts, J.B.: First-passage time for oscillators with nonlinear damping, ASME J. App. Mechs., 45(1978), 175-180.

42. Roberts, J.B.: First passage probability for nonlinear oscillators, ASCE J. Engng. Mechs. Div., 102(1976), 851-866.

43. Roberts, J.B.: First-passage time for randomly excited nonlinear oscillators, J.Sound Vib., 109(1986), 33-50.

44. Spanos, P.D. and G.P. Solomos: Barrier crossing due to transient excitation, ASCE J. Engng. Mechs., 110(1984), 20-36.

45. Roberts, J.B.: An approach to the first-passage problem in random vibration, J.Sound Vib., 8(1968), 301-328.

46. Roberts, J.B.: Probability of first passage failure for stationary random vibration, AIAA J., 12(1974), 1636-1643.

Chapter 2

SAFETY INDEX, STOCHASTIC FINITE ELEMENTS AND EXPERT SYSTEMS

F. Casciati
University of Pavia, Pavia, Italy

Abstract

The material presented in this Chapter has a two-fold objective. Firstly, it introduces those concepts in the Theory of Reliability which are useful in the Mechanics of Solids and Structures. The goal, here, is to link the general theoretical statements of Chapter 1 with the methods and concepts of the following two Chapters and, hence, with the subsequent specialistic applications.
Secondly, the present Chapter provides a short presentation of two research fields which are provoking a growing interest amongst practitioners, namely "Stochastic Finite Elements" (with their potentialities for dealing with complex mechanical systems) and "Expert Systems". The application of the latter approach is mainly in monitoring and upgrading existing plants and buildings.

Preface

It is always embarrassing to write a text of lecture notes. One presents results that the scientific community (and oneself and/or coworkers, as members of it) have obtained in past years. To be honest, one cannot forget that some colleagues have devoted their time to provide the correct systematization of the matter one is illustrating. One is just acting as a filter with the aim of balancing the readability of the result with the limited space that a lecture note book provides. One can only build a grid where the single nodes are the property of the scientific community.

For these lecture notes some references are of basic importance. They are listed in this preface to emphasize their intrinsic merit. This chapter is a short summary of Ref. [1] to which the reader is referred for further details, as well as to Ref.[2] for the bases of probabilistic methods in Structural Engineering. Refs. [3] to [7] also had a special importance in preparing these notes.

2.1 SAFETY INDEX AND FAST INTEGRATION RELIABILITY METHOD

2.1.1 Definitions

According to Ref. [3], the "structural performance of a whole structure or part of it should be described with reference to a specified set of adverse states beyond which the structure no longer satisfies the performance requirements. Each adverse state is the boundary of an adverse event. A binary description of the performance is inherent in the adverse event concept".

A list of adverse events (limit states) is given below:

- loss of static equilibrium of the structure, or a part of the structure, considered as a rigid body,
- localized rupture of critical sections of the structure caused by exceeding the ultimate strength (possibly reduced by repeated loading), or the ultimate deformation of the material,
- transformation of the structure into a mechanism,
- general or local instability,
- progressive collapse,
- deformation which affects the efficient use or appearance of structural or non-structural elements,
- excessive vibration, which can cause discomfort and/or alarm
- local damage (including cracking), which affects the durability or the efficiency of the structure.

The uncertainties of the mechanical models and their parameters can be introduced by presenting them in terms of concepts from mathematical probability theory. According to Ref. [3], some of the parameters of relevance are presented as being basic variables "in the sense that they are assumed to carry the entire input information to the mechanical model".

The basic variables are material parameters, external action parameters, and geometrical parameters. All other quantities are functions of these basic variables. They are cross section resistances, member buckling resistances, load effects, areas, volumes, safety margins, event margins, etc.

The basic variables may be joined into a finite-dimensional vector. Then the uncertainties of the problems are modeled by letting this vector be a vector of random variables. Mutual stochastic dependence between the components of this random vector may occur.

Basic variables may more generally be functions in time and space. The corresponding probabilistic concept is that of a random process, or a random field.

Uncertainties from all essential sources must be evaluated and integrated into the reliability model. Types of uncertainty to be taken into account are, according to Ref. [3], physical (intrinsic) uncertainty, statistical uncertainty, and model uncertainty.

"Physical uncertainty is the ubiquitous background randomness, the level of which may or may not be controlled by active means. Statistical uncertainty is due to limited information as it is provided by a sample of finite size. Model uncertainty is due to the necessary idealizations on which the physical model formulation and the distributional model formulation are based" [3]. The last type of uncertainty may, for each adverse event, be described as uncertainty of the corresponding adverse state surface.

Statistical uncertainty must be quantified when a small sample of observations of a basic variable is available. By use of a non-informative prior, a predictive posterior distribution is calculated [2][8]. This distribution must be applied in a reliability analysis. (The posterior density is obtained as being proportional to the product of the likelihood function and the prior density (according to Bayes' formula) The likelihood function is defined by the joint distribution of the basic variables $\underline{X}_1, \ldots, \underline{X}_N$ considered as a function of the parameters.).

The reliability model must be formulated in such a way that, for each given adverse state surface, a judgemental random vector $\{\underline{J}\} = (\underline{J}_1, \ldots, \underline{J}_N)^T$ can be associated with the basic random vector $\{\underline{X}\} = \{\underline{X}_1, \ldots, \underline{X}_N\}^T$. The resulting random vector $\{\underline{X}_J\}$ replaces $\{\underline{X}\}$ in the reliability calculation (here an underlined letter denotes a random variable; vectors and matrices are denoted by {.} and [.] respectively).

An event margin for a given event is any function of the basic variables with the property that it takes a negative value if, and only if, the event occurs. When referred to the adverse state, it is denoted as the safety margin.

"Information becoming available after the design of a structure can be formulated in a framework of event margins. This additional information can be utilized in reliability updating" (see Section 2.4).

Decision theory can be applied in order to obtain optimal target reliability levels. It is required, however, that the intangible part of the losses due to failure be separated from the cost of failure. For this purpose multiobjective mathematical programming should be adopted. The set of the best solutions (Pareto optima) are the ones which maximize the reliability for given values of the constructional cost.

Among these solutions the one to be adopted is selected as a compromise between financial resources and ethical considerations [9].

Required minimal reliability levels make sense only within the specification of a reference period. The reference period should generally equal the anticipated lifetime of the structure (e.g. 50 years, or 100 years for special facilities) (see Chapter 4).

2.1.2 Time invariant reliability assessment

Assume that the state of a structural component (safe or fail) depends on the vector $\{\underline{X}\} = \{\underline{X}_1, \underline{X}_2, \ldots, \underline{X}_N\}^T$ of N time-invariant uncertain basic-variables. Let $p_{\{\underline{X}\}}(\{X\})$ be the joint probability density function (JPDF) of the basic variables. The associated cumulative distribution is defined by

$$P_{\{\underline{X}\}}(\{X\}) = \int_\Theta p_{\{\underline{X}\}}(\{r\})\ \{dr\} = \text{Prob}\ [\bigcap_{i=1}^{N} \{\underline{X}_i \leq X_i\}] \tag{2.1}$$

where Θ is the region where $r_i \leq X_i$, for all i.

Each set of values of the basic variables corresponds to one of the two states of the component. The surface $\partial D_{\{X\}}$ dividing the {X}-space into a safe region, D, and a failure region, U, is called the failure surface.

A function g({X}) is a safety margin (or "limit state function" or "performance function") if:

- g({X}) > 0 denotes safe states, i.e. $\{X\} \in D$;
- g({X}) = 0 denotes limit states, i.e. $\{X\} \in \partial D_{\{X\}}$;
- g({X}) < 0 denotes failure states, i.e. $\{X\} \in U$.

The probability that $\{\underline{X}\}$ falls into the failure region U = {g({X})<0} is the probability of failure. It is given by:

$$P_f = \text{Prob}\ [\{\underline{X}\} \in U] = \int_U p_{\{\underline{X}\}}(\{X\})\ \{dX\} \tag{2.2}$$

The failure probability can be evaluated by integration and simulation methods (see Chapter 3). It can also be estimated by "second moment" methods. Their characteristic is that the actual JPDF is approximated by a Gaussian model.

Let $\{\underline{X}\}$ be a vector of independent normal basic variables. Let the failure function g({X}) be linear [1]. Transform the vector $\{\underline{X}\}$ into a standard normal vector $\{\underline{Z}\} = \{\underline{X}\}^* = \{\underline{X}_1^*, \underline{X}_2^*, \ldots, \underline{X}_N^*\}^T$ (having a normal distribution with zero mean $\{\mu_{\{\underline{Z}\}}\}$ and unit variance $\{s_{\{\underline{Z}\}}\}$), defined as

$$\underline{Z}_i = (\underline{X}_i - \mu_{\underline{X}i}) / s_{\underline{X}i} \tag{2.3}$$

The linear failure function is mapped into the {Z}-space as:

$$g_{\{Z\}}(\{Z\}) = C_0 + \sum_{i=1}^{N} C_i\ Z_i \tag{2.4}$$

The failure domain U is the half-space $g_{\{Z\}}(\{Z\}) \leq 0$ of the equation

$$c_0 / (\Sigma_i c_i^2)^{1/2} + \sum_{i=1}^{N} c_i z_i / (\Sigma_i c_i^2)^{1/2} \leq 0 \qquad (2.5)$$

Appropriate notations lead one to write

$$\beta + \{\alpha\}^T \{Z\} = \beta + y \leq 0 \qquad (2.6)$$

In equation (2.6) the scalar y is the product $\{\alpha\}^T\{Z\}$. The corresponding $\underline{y} = \{\alpha\}^T\{\underline{Z}\}$ is again a standard normal variable ($E[\underline{y}] = 0$; $Var[\underline{y}] = 1$), due to the linear nature of the transformation.
The failure probability is given by

$$P_f = \text{Prob}\,[g_{\{Z\}}(\{\underline{Z}\}) \leq 0 = \text{Prob}\,[\beta + \{\alpha\}^T \{\underline{Z}\} \leq 0]$$

$$= \text{Prob}\,[\beta + \underline{y} \leq 0] = \text{Prob}\,[\underline{y} \leq -\beta] = \Phi(-\beta) \qquad (2.7)$$

where $\Phi(-\beta)$ is the standard normal integral computed at $-\beta$ (see Table I).
The term β is called the "safety index" and is given by

$$\beta = E[g(\{\underline{Z}\})]/(Var\,[g(\{\underline{Z}\})])^{1/2} \qquad (2.8)$$

It was first suggested as a measure of reliability in Ref.[10]. Cornell introduced the form (2.8) in 1969 [11].

Table I - Values of β versus the standard normal integral $\Phi[-\beta]$ (from Ref.[1]).

ß	1.5	2.0	2.32	2.5	3.0
Φ[-ß]	$6.6\ 10^{-2}$	$2.2\ 10^{-2}$	$1.0\ 10^{-2}$	$6.2\ 10^{-3}$	$1.3\ 10^{-3}$
ß	3.09	3.5	3.72	4.0	4.27
Φ[-ß]	$1.0\ 10^{-3}$	$2.1\ 10^{-4}$	$1.0\ 10^{-4}$	$3.2\ 10^{-5}$	$1.0\ 10^{-5}$
ß	4.5	4.75	5.0	5.2	5.61
Φ[-ß]	$3.3\ 10^{-6}$	$1.0\ 10^{-6}$	$2.9\ 10^{-7}$	$1.0\ 10^{-7}$	$1.0\ 10^{-8}$

Since the vector $\{\alpha\}$ contains the direction cosines of the normal to the plane $\beta + \alpha^T\{Z\} = 0$ in equation (2.7), β is its distance from the origin. Therefore, the safety index β corresponds to the shortest distance from the origin to the limit state surface $g_{\{Z\}}(\{Z\}) = 0$ in the $\{Z\}$-space.

This ß approach can still be applied by linearizing $\partial D_{\{Z\}}$ in certain points (ß points), when the failure surface D_Z is not linear. A truncated Taylor series expansion can be used for this purpose.
An alternative linearization scheme is achieved by introducing the first-order second-moment reliability index β_{HL} defined by Hasofer and Lind [12].

Let $\{\underline{X}\}$ be a basic vector. It is mapped into a normalized and uncorrelated vector $\{\underline{Z}\}$, with $E[\{\underline{Z}\}] = 0$ and Cov $[\{\underline{Z}\}\{\underline{Z}\}^T] = [\Sigma_{\{\underline{Z}\}}] = [I]$, by a transformation defined as

$$\{\underline{Z}\} = [U_I]^T (\{\underline{X}\} - E[\{\underline{X}\}]) \tag{2.9}$$

and

$$[\Sigma_{\{\underline{Z}\}}] = [U_I]^T [\Sigma_{\{\underline{X}\}}] [U_I] = [I] \tag{2.10}$$

Due to the rotational symmetry of the second moment information on the $\{\underline{Z}\}$ variables, the distance to the failure surface can be calculated [13] by the expression (Fig.2.1)

$$\beta(\{Z\}) = (\{Z\}^T\{Z\})^{1/2} \quad ; \quad (\{Z\} | g_{\{Z\}}(\{Z\})) = 0\} \tag{2.11}$$

in the {Z}-space and by

$$\beta(\{X\}) = [\{X\} - E[\{\underline{X}\}])^T [\Sigma_{\{\underline{X}\}}]^{-1} (\{X\} - E[\{\underline{X}\}])]^{1/2} \quad ; \quad (\{X\} | g(\{X\}) = 0) \tag{2.12}$$

in the {X}-space.

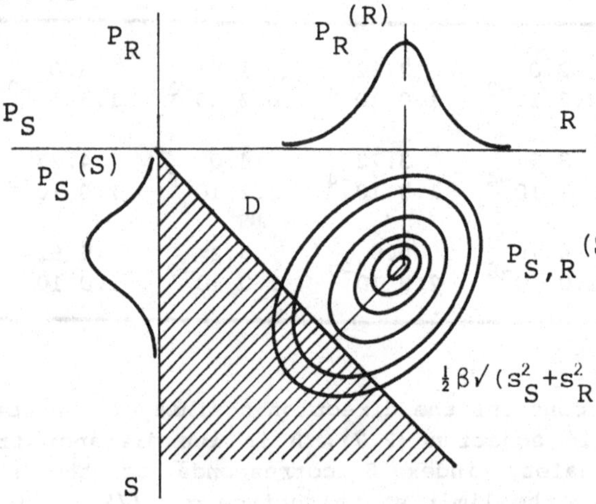

Fig. 2.1 - The distance to the failure surface as safety index

The reliability index proposed by Hasofer and Lind is, eventually, the solution of the linear optimization problem

$$\beta_{HL} = \min \left[(\{X\} - E[\{\underline{X}\}])^T \, [\Sigma_{\{\underline{X}\}}]^{-1} \, (\{X\}-E[\{\underline{X}\}]) \right]^{\frac{1}{2}} \tag{2.13}$$

under the constraint $(\{X\} | g(\{X\}) = 0)$.
β_{HL} is the smallest distance from the origin of the {X}-space to the point on the failure surface. The particular point {X} which satisfies equation (2.13) is called the "design point" and is denoted $\{X\}^x$. It is the most likely failure point, according to Ref. [14]. In fact, the JPDF of independent multinormal variables is invariant against rotations and, hence, its maximum along $g(\{X\}) = 0$ is reached at $\{X\}^x$.

The constrained minimization in Eq.(2.13) can be regarded as an unconstrained optimization problem by introducing a Lagrangian multiplier [15]. For complex structural systems, however, the evaluation of β_{HL} must be pursued by numerical approaches.
When the constraint $g(\{X\}) = 0$ is linear, the problem becomes a quadratic programming problem and optimization algorithms can be employed [16]. These techniques can be used also when the constraint $g(\{X\}) = 0$ is nonlinear and a feasible vector $\{X\}_f$ (i.e. $g(\{X\}_f) = 0$) is known. Otherwise, algorithms which solve nonlinear minimization problems under nonlinear constraints must be applied [17].
Many other techniques are available to solve such an optimization problem. In particular, the iterative algorithm by Rackwitz and Fiessler [18] was proposed, exactly for this purpose, for cases where efficient general-purpose algorithms are not available (see Chapter 3).

When the failure surface $g_{\{Z\}}(\{Z\})$ is linear, $\Phi(-\beta_{HL})$ gives the probability of failure P_f.
If the failure surface is not linear, a linearization can be introduced by the tangent hyperplane at the design point $(\{X\}^x)$ ("first-order-reliability-method" = FORM). In this case, the relationship between the failure probability P_f and β_{HL} is

$$0 \leq P_f \leq \Phi(-\beta_{HL}) \tag{2.14}$$

if $g_{\{Z\}}(\{Z\})$ is convex to the origin. $\Phi(.)$ denotes the standardized Gaussian CDF. By contrast, if $g_{\{Z\}}(\{Z\})$ is concave, one has

$$\Phi(-\beta_{HL}) \leq P_f \leq 1 - \chi_N^2(\beta_{HL}^2) \tag{2.15}$$

where $\chi_N^2(.)$ is the χ^2 distribution with N degrees of freedom ($p_{\underline{q}}(q) = q^{(N/2)-1} e^{-q/2}/[2^{N/2} \Gamma(N/2)]$, with $E[\underline{q}] = N$ and $Var[\underline{q}] = N^2$). Remember that N is the number of the basic variables in $\{\underline{Z}\}$.

If a linearization appears inappropriate, a more accurate approximation of the nonlinear failure surface can be obtained by introducing a hyperparaboloid, fitted at the design point ("second-order-reliability-method" = SORM). In this case the probability of failure can be estimated from the relations given in Refs. [19] and [20].

In general the basic variables $\{\underline{X}\}$ are not normally distributed and not mutually independent. A suitable transformation of the vector $\{\underline{X}\}$ into a set of random variables $\{\underline{Z}\}$, which are uncorrelated, standardized and normally distributed, has been given in [21], by using the Rosenblatt transformation.
By the Rosenblatt transformation one makes use of probability information on the variable $\{\underline{X}\}$. However the reliability measure is still based on the second moments of the state variables. Improvements can be pursued by using an approximation of the higher moments. Grigoriu [22] proposed a weighted system of second moment functions. Winterstein and Bjerager [23] suggest exploiting information on the third and fourth moments: in this way one greatly improves accuracy in the distribution tails; it is these tails which govern reliability (see Chapter 3).

2.1.3 Load combination

If a multivariate load condition includes time varying stochastic loads, some circumstances make the assessment of the reliability difficult.

1) Each varying-in-time load is generally modelled independently of the other possible concurrent actions. It follows that the actual input acting on the structural system is rarely known, as a stochastic-mechanics problem requires, i.e. by the knowledge of the properties of the single processes and by the relevant cross-properties.

2) The concept of a "more unfavourable level"is strictly related, for a single action, to either the maximum or the minimum value of its intensity. By contrast a more unfavourable load combination depends on the shape of the boundary of the safety domain and, hence, it is different from limit state to limit state, i.e. the problem is highly dependent on the structure.

3) Even if special assumptions are made, in order to remove the previous two difficulties, the available stochastic approaches are often inadequate. This is not the case (however) for structures which are assumed to behave in a linear elastic way (linear boundary of the safety domain) for which the total load can be expressed by a linear function of the load intensities (input combination).

State-of-the-art papers on load combination have been presented at some recent international conferences ([21][25] and [26] among others) and many recent books on structural reliability, or stochastic mechanics, devote one or more sections to the problem [27][1][28].

Technical problems in which the theoretical difficulties are predominant on the operative aspect give rise, generally, to two parallel approaches. The first one tackles the problem in its more general formulation and attempts to solve it in a general way, or under suitable assumptions. The second approach neglects the central problem and attempts to achieve technical results by introducing simplified models or bounding techniques. This general trend has caused a diversificaton of interests amongst researchers working in the load combination area, whose results can be grouped into the following four categories:

1) approaches which pursue the solution of the problem in its general formulation (see next sub-section);

2) approaches which pursue the rigorous solution of a simplified problem of common practical interest. A basic procedure of structural analysis is founded on the assumptions of linear structural behaviour and a linear combination of the load intensity variables, or their (linear or nonlinear) functions. Then, the load combination problem can be formulated as follows: given n loads $\underline{W}_i$, i = 1,...,n, which are modelled by stationary independent stochastic processes, derive the stochastic properties of the process

$$\underline{W}(t) = \sum_{i=1}^{n} \underline{W}_i(t) \qquad (2.16)$$

or

$$\underline{R}(t) = \sum_{i=1}^{n} c_i \underline{L}_i(t) \qquad (2.17)$$

where $\underline{L}_i(t)$ is a scalar funtion of the load process $\underline{W}_i$ and c_i is a structural influence coefficient. This enables one to assess the probability that $\underline{W}(t)$ or $\underline{R}(t)$ crosses, with positive slope (up-crosses) the relevant given barrier during a given reference period (see Refs. [29] and [30] among others). The problem is then to characterize the maximum of this new process (see Chapter 1).

3) proposals of approximate models, enabling one to pursue general results. These approaches have been all conceived in view of a reliability analysis and consist of three main steps:
1) select a finite set of constant-in-time load vectors, which care all possible unfavourable situations;
2) solve the time independent reliability in terms of the values calculated at step 1).

Three load combination models are widely employed in structural reliability analysis: i) Turkstra's rule [28]; ii) the Ferry-Borges-Castanheta model [29] and iii) the load coincidence model, proposed by

Wen [37] and adopted by several researchers [34][35][36]. The reader is referred to Ref. [25] for more details.

4) translation of the previous results into common rules of practice (see Chapter 4)

2.1.4 Structural reliability under dynamic excitation

It is known (see Chapter 1) that the safety of a deterministic structure under one stochastic load can be evaluated by solving a barrier crossing problem, the barrier being the load carrying capacity of the structure (Fig.2.1a and c). However, even if the hypothesis of a deterministic structure is accepted, the problem has only been solved, in the univariate scheme, for very particular cases.

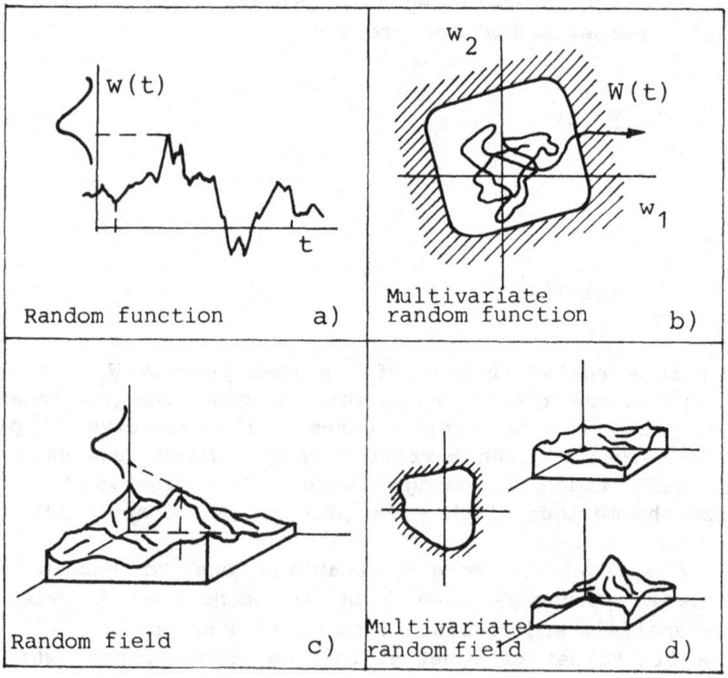

Fig.2.2 - First excursion problem for univariate and multivariate random process and field.

For multi-excitation, or for random structural parameters, the problem must be formulated in a different way [4].

Consider a system depending on a N-dimensional vector of uncertain time-invariant parameters $\{\underline{X}\}$. When the system is subjected to a dynamic stochastic excitation, it is considered to perform satisfactorily if a response vector $\{\underline{R}(t)\}$, $t \geq 0$, does not leave a domain D during a reference period $(0, \tau)$. The probability of the event

$$P_S(\tau) = \text{Prob}(\underline{R}(t) \in D, 0 \leq t < \tau), \tag{2.18}$$

is referred to as the system reliability. In general it cannot be determined exactly, even for deterministic systems (i.e., when $\{\underline{X}\}$ is equal to a particular value $\{X\}$), or for an excitation vector constant in time (i.e. when the loads are random variables).
However, it can be estimated from the mean outcrossing rate, or the first passage time of $(\{\underline{R}(t)\} | \{\underline{X}\} = \{X\})$ relative to set D (see Refs.[37][38] and Chapter 1 of this book). The response of a deterministic system with $\{\underline{X}\} = \{X\}$ can be determined by the classical methods of random vibration (see Refs. [39][48][14] and Chapter 1).
Reliability analysis of uncertain systems is quite involved because classical methods of random vibration cannot be applied directly. A two-phase method can be used to calculate $P_S(\tau)$ [4]:
I) one computes the conditional reliabilities $P_S(\tau | \{\underline{X}\} = \{X\})$ corresponding to deterministic systems with $\{\underline{X}\} = \{X\}$; this is done by the classical methods of random vibration and mean D-outcrossing rates or D-first passage times of $(\{\underline{R}(t)\} | \{\underline{X}\} = \{X\})$;
II) an unconditional reliability $P_S(\tau)$ is determined from

$$P_S(\tau) = \int P_S(\tau | \{\underline{X}\} = \{X\})\ p_{\{\underline{X}\}}(\{X\})\ \{dX\} \tag{2.19}$$

where $p_{\{\underline{X}\}}(\{X\})$ is the probability density function of $\{\underline{X}\}$. Since the knowledge of $P_S(\tau | \{\underline{X}\} = \{X\})$ is required, for all the values of $\{X\}$, the integration in equation (2.19) is not easy to carry out. It can be determined approximately by discretizing the space of uncertain parameters or by simulating samples of $\{\underline{X}\}$ and developing estimates of $P_S(\tau)$ based on these samples. However, the efficiency and/or accuracy of these approximations may not be satisfactory. First or second order reliability methods have been introduced to improve efficiency, as will be shown in the following.

An alternative two-phase method can be employed when a single stochastic excitation, of random intensity, $\underline{I}$, acts on the structure. In this case

$$P_S(\tau) = \int P_S(\tau | \underline{I}=I)\ p_{\underline{I}}(I)\ dI \tag{2.20}$$

where the integral is no longer a multidimensional integral. The plot of the term $P_S(\tau|\underline{I}=I)$ is often referred to as a fragility curve [1]. The approach summarized in equation (2.20) becomes feasable in the following intermediate case:

1) the system is deterministic in terms of geometry, stiffness and damping. This means that the structural response to a stochastic excitation can be estimated by applying random vibration algorithms.
2) the system randomness is concentrated into the parameters by which D is defined: for instance, the parameters of the local failure criterion in a series system scheme. In this case one has

$$P_s(\tau|\underline{I}=I) = \pi_i \int_0^\infty P_{\underline{R}i}(R)\, p_{\underline{R}i*}(R)\, dR \tag{2.21}$$

where $\underline{R}_i$ is the random local response in the i-th series element and $\underline{R}^*_i$ is the relevant strength.

When assumption 1) is not satisfied, stochastic finite element algorithms [1] must be adopted in order to establish the probabilistic features of the response (see Section 2.2)

The differential equations for the mean and covariance function of linear systems can be used to determine the second-moment characteristics of $(\{\underline{R}(t)\}|\{\underline{X}\}=\{X\})$. These equations can be applied directly to the equations of motion of a dynamic system, or to an alternative form of it, expressed in a new system of coordinates defined by the eigenvectors of the system matrix. The latter approach allows the determination of the mean and covariance functions of the response in a closed form, even when the excitation is not a white noise [42] [43].
The means and covariance functions of the conditional response $(\{\underline{R}(t)\}|\{\underline{X}\}=\{X\})$ can be derived simply from the corresponding functions of the motion parameters, because these processes are linearly related. Results of Chapter 1 (see Refs. [37] and [38]) can then be used to calculate the mean D-outcrossing rates $\nu(t;\{X\})$ of $(\{(\underline{R}(t)\}|\{\underline{X}\}=\{X\})$. Alternatively use of equation (2.20) can be pursued.

By contrast, there is no practical method for finding, exactly, the conditional probability of the response $(\{\underline{u}(t)\}|\{\underline{X}\}=\{X\})$ in the nonlinear case. However, several approximate techniques have been developed for calculating the moments, and other probabilistic characteristics of $(\{\underline{u}(t)\}|\{\underline{X}\}=\{X\})$, e.g., equivalent linearization, perturbation, moment closure, and simulation [44]. The equivalent linearization method [41] is extensively used in the analysis of realistic nonlinear systems [45 to 52]. It approximates the actual equations of motion by linear differential equations that are equivalent, in some sense, to the original one. This is done by preserving the hysteretic nature of the material (finite-element) constitutive law. Thus, the theory of linear systems can be applied to characterize $(\{\underline{u}(t)\}|\{\underline{X}\}=\{X\})$ approximately. However, the methods assume incorrectly that this process follows a Gaussian probability law. This assumption can result in unsatisfactory estimates of conditional reliability $P_S(\tau|\{\underline{X}\}=\{X\})$.

The moment closure method is also used for the random vibration analysis of nonlinear systems. It is based on differential equations that can be developed for the conditional process $(\{\underline{u}\}(t)\}|\{\underline{X}\}=\{X\})$ by, e.g., the Itô differential rule. The moment equations constitute an infinite sequence that cannot be solved exactly. Heuristic approximations relating higher to lower order moments of the response process have been used for solution purposes. These approximations are referred to as closure techniques. The resultant moments of $(\{\underline{u}(t)\}|\{\underline{X}\}=\{X\})$ can be used to develop approximations to the conditional reliability $P_S(\tau|\{\underline{X}\}=\{X\})$.

2.1.5 Fast integration reliability analysis

A method for the reliability analysis of uncertain systems subject to stationary and nonstationary excitations, is based on the use of first/second order methods, random vibration techniques and the crossing theory of random processes. It is commonly referred to as "fast integration " or "nested" reliability analysis [53].
Consider a response $\{\underline{R}(t)\}$ depending on a random vector $\{X\}$ and a safe set D. Let $\nu(t;\{\underline{X}\}=\{X\})$ and $T(\{\underline{X}\}=\{X\})$ be the mean D-outcrossing rate at time $t\geq 0$ and the D-first passage time of conditional process $\{\underline{R}(t)\}|\{\underline{X}\}=\{X\})$. The probability that $\{\underline{R}(t)\}|\{\underline{X}\}=\{X\})$ does not leave domain D in (0,t) is $P(T\{\underline{X}\}=\{X\})>t)$ and is approximated by (Chapter 1)

$$P_S(\tau|\{\underline{X}\}=\{X\}) \cong \exp[-\int_0^{\tau} \nu(t,\{\underline{X}\}=\{X\})\, dt] \tag{2.22}$$

provided that $\{\underline{R}(0)\} \in D$ with nearly unit probability.
Let $\{Z\}=\{h(\{X\})\}$ be a nonlinear transformation mapping vector of the uncertain parameters $\{\underline{X}\}$ onto an N-dimensional space of zero mean, unit variance, independent Gaussian variables $\{\underline{Z}\}^T=\{\underline{Z}_1,\ldots,\underline{Z}_N\}$. Consider an additional zero mean unit variance independent Gaussian variable $\underline{Z}_{N+1}$ and a "safe" set D* in space $(\{Z\},Z_{N+1})$ defined by the condition

$$g(\{Z\},Z_{N+1}) = Z_{N+1} \cdot \Phi^{-1}(P_s(\tau|\{h^{-1}(\{Z\})\})) < 0 \tag{2.23}$$

where Φ = the distribution of the standard Gaussian variable. It can be shown that the probability content of D* coincides with unconditional probability $P_S(\tau)$ [53]. Thus, FORM/SORM algorithms can be applied in the (N+1)-dimensional standard Gaussian space $(\{Z\},Z_{N+1})$ for the safe set D* to calculate the reliability $P_S(\tau)$. Note that the evaluation of function $g(\{Z\}, Z_{N+1})$ in equation (2.23) requires a random vibration analysis of a deterministic system with parameters $\{X\}=\{h^{-1}(\{Z\})\}$ (see Chapters 3 and 5 for more details on this procedure).

2.2 STOCHASTIC FINITE ELEMENTS

Finite-element techniques are presently the best analysis approach for any complex structural or mechanical system [1]. The level of sophistication that linear and nonlinear analyses require is quite different. Randomness can affect the external actions, the system properties or both of them. An intermediate objective, for successive reliability assessment is to determine the probabilistic properties of the structural response. A special response quantity is, in particular, the performance function $g(\{X\})$.

2.2.1 Linear and nonlinear deterministic systems under random actions

The simplest input-output numerical relationship is written for a linear deterministic system with static random loads:

$$[K] \quad \{\underline{u}\} = \{\underline{W}\} \tag{2.24}$$

where $[K]$ is the deterministic stiffness matrix, $\{\underline{W}\}$ is a vector of random loads and $\{\underline{u}\}$ is a vector of generalized displacements. The randomness of $\{\underline{W}\}$ causes the $\{\underline{u}\}$ to be random.
It is well known [2], then, that the average vector $\{\mu_{\{\underline{u}\}}\}$ and the covariance matrix $[\Sigma_{\{\underline{u}\}}]$ of the displacements depend only on the average vector $\{\mu_{\{\underline{W}\}}\}$ and the covariance matrix $[\Sigma_{\{\underline{W}\}}]$ of the loads:

$$\{\mu_{\{\underline{u}\}}\} = [K]^{-1} \{\mu_{\{\underline{W}\}}\} \tag{2.25}$$

$$[\Sigma_{\{\underline{u}\}}] = [K]^{-1} \; [\Sigma_{\{\underline{W}\}}] \; ([K]^{-1})^T \tag{2.26}$$

For loads $\{\underline{W}\}$ which are jointly Gaussian, the linear nature of the system means that the displacements are also jointly Gaussian, and the problem, in the derived variables, is fully solved.

Analogously, if the system has a transfer function $[H(\omega)]$ and the loads are stationary stochastic processes, one can introduce the matrix $[G_{\{\underline{W}\}}(\omega)]$ whose elements are the Fourier transforms of the autocorrelation functions (G_{ii}) and of the cross-correlation functions (G_{ij} , $i \neq j$). The corresponding matrix for $\{\underline{u}\}$ is given by [2]:

$$[G_{\{\underline{u}\}}(\omega)] = [H(\omega)]^2 \, [G_{\{\underline{W}\}}(\omega)] \tag{2.27}$$

Again, for jointly Gaussian stochastic load processes, the problem is fully solved by equation (2.27) and

$$\{\mu_{\{\underline{u}\}}\} = (\{\mu_{\{\underline{W}\}}\}/2\pi) \int_{-\infty}^{\infty} d\omega \int_{-\infty}^{\infty} [H(\omega)] e^{-i\omega t} \, dt \tag{2.28}$$

Equations (2.25) to (2.28) show clearly that the basic task here is structural discretization i.e. determination of the structural matrices $[K]$ and $[H]$.

Nonlinearity can be introduced by either the presence of large displacements and deformations or through nonlinearities in the material constitutive law. In the latter case, for a structure discretized into finite elements, one expresses the generalized displacement $\{u(\{x\})\}$ in the i-th element in terms of the nodal displacements $\{u_i\}$ by a matrix of shape functions $[\Lambda_i(\{x\})]$:

$$\{u(\{x\})\} = [\Lambda_i(\{x\})]\ \{u_i\} \tag{2.29}$$

The compatibility relation provides then the deformation vector $\{\epsilon(\{x\})\}$ ($\underline{B}$ = compatibility matrix)

$$\{\epsilon(\{x\})\} = [B]\ \ \{u\} \tag{2.30}$$

and the constitutive law has the incremental form ($[D]$ = elastic matrix)

$$\{\dot{\sigma}(\{x\})\} = [D](\{\dot{\epsilon}\} - \{\dot{\epsilon}_p\}) \tag{2.31}$$

where $\{\epsilon_p\}$ denotes the vector of the inelastic deformations which act as distortions in the determination of the stress vector $\{\sigma(\{x\})\}$.
The general form of the equilibrium equation is then:

$$\Sigma_i(\int_{V_i} [B_i]^T\ [D_i]\ [B_i]\ dV)(\{du_i\} + c\{d\dot{u}_i\}) +$$

$$(\int_{V_i} \rho[\Lambda_i][\Lambda_i]^T\, dV)\ \{d\ddot{u}_i\} \quad - (\int_{V_i} [B_i]^T\ [D_i]\{d\epsilon_{ip}\}\, dV) +$$

$$- (\int_{V_i} [\Lambda_i]^T\ \{dg\}\, dV) - (\int_{V_i} [\Lambda_i]^T\{df\}\, dA) = 0 \tag{2.32}$$

where ρ is the density, $\{g\}$ denotes the body forces and $\{f\}$ the surface forces. The coefficient of the velocity $\{du\}$ due to the damping conditions is assumed to be proportional to the stiffness by a factor c in equation (2.32).In a more compact form, one writes.

$$[K_i]\{du_i\} + c[K_i]\{du_i\} + [m_i]\{d\ddot{u}_i\} - \{dW_i\} - \{dW_i^{P}\} = 0 \tag{2.32}$$

Therefore, the term introducing the nonlinearity is $\{dW_i^P\}$ linearly related with $\{d\epsilon_{ip}\}$at the integration Gauss points. At each time it is a function of the previous time history since the material constitutive law is not reversible.

When $\{\underline{W}\}$ is a vector of stochastic processes, equation (2.33) is satisfactorily treated by stochastic equivalent linearization techniques [1][51]. The reader is referred to Chapters 1 and 5 for further details.

2.2.2 General random systems

For linear systems characterized by random properties and subjected to random actions, a common unified approach is not available in the literature. Early procedures are provided in Ref.[54] to [57] and [2].
The structural response is obtained by using a finite element algorithm. The data ({X}) are just a realization of a random vector ({X}) which can be regarded as a discretization of several random fields [58][59]. This leads to a stochastic finite-elements formulation, especially when the numerical experiments to be carried out are not fully randomized but are planned according to design theory. One can distinguish, in this field, studies aimed at assessing system reliability directly [60] and analyses of uncertainty propagation.
Der Kiureghian [60] has investigated the assessment of reliability by level-2 reliability methods. For this purpose, he expressed the performance function g(.) as a function of the vector {r} of resistances and of the vector {s} of load effects:

$$\{s\} = [q]^T \{u\} = [q]^T [K]^{-1} \{W\} \tag{2.34}$$

The latter ones must be compared with {r}. The derivatives of g({r},{s}) at the linearization point $\{Z^x\}$ in the space of the standardized design variables {Z} are then determined by the algorithm which minimizes the distance from the origin.
They are written in [61] as:

$$\{\partial g/\partial Z_j\}_{\{Z\}=\{Z^x\}} = \{(\partial g/\{\partial r\})^T \cdot \{\partial r\}/\partial Z_j + (\partial g/\{\partial s\}^T \cdot \{\partial s\}/\{\partial Z_j\}\}_{\{Z\}=\{Z^x\},\{r\}=\{r^x\},\{s\}=\{s^x\}} \tag{2.35}$$

and

$$[\{\partial s\}/\partial Z_j] = \{([\partial q]^T/\partial Z_j)\ \{u\} + [q]^T[K]^{-1}(-\partial[K]/\partial Z_j + \partial\{W\}/\partial Z_j\}_{\{\underline{Z}\}=\{Z^x\}} \tag{2.36}$$

Looking to the solution of the uncertainty propagation problem, the analysis of linear systems can take advantage of special devices, such as second order perturbation [62][63] or Neumann expansions [65].
The perturbation method was used by Nakagiri [62] [64] and his coworkers, writing the single design variable in the form

$$\underline{X}_j = X_j\ (1 + \underline{\alpha}_j) \tag{2.37}$$

The Taylor expansion of the stiffness matrix is

$$[\underline{K}] = [\bar{K}] + \sum_{j=1}^{N} [K_j^{I}]\ \underline{\alpha}_j + (1/2) \sum_{i=1}^{N} \sum_{j=1}^{N} [K_{ij}^{II}]\underline{\alpha}_i\ \underline{\alpha}_j \tag{2.38}$$

where $[K_i^{I}]$ and $[K_{ij}^{II}]$ must be obtained by differentiation.

The Taylor series for {u} is then:

$$\{\underline{u}\} = \{\bar{u}\} + \sum_{j=1}^{N} \{u_j^{T}\}\underline{\alpha}_j + (1/2)\sum_{i=1}^{N}\sum_{j=1}^{N} \{u_{ij}^{II}\}\underline{\alpha}_i\,\underline{\alpha}_j \tag{2.39}$$

whose coefficients are determined for deterministic {W} from the set of equations

$$[\bar{K}]\,\{u\} = \{W\}$$

$$[\bar{K}]\,\{u_i\}^{T} = -[K_i^{I}]\,\{\bar{u}\} \tag{2.40}$$

$$[\bar{K}]\,\{u_{ij}^{II}\} = [K_i^{I}]\{u_j^{I}\} - [K_j^{I}]\{u_j^{I}\} - [K_{ij}^{II}]\{\bar{u}\}$$

The second order approximations of the mean and variance of {u} are eventually found to be:

$$E[\{\underline{u}\}] = \{\bar{u}\} + (1/2)\,\Sigma_i\Sigma_j\{u_{ij}^{II}\}\,E[\underline{\alpha}_i\,\underline{\alpha}_j] \tag{2.41}$$

$$Var[\{\underline{u}\}] = \Sigma_i\,\Sigma_j\,\{u_i^{I}\}\{u_j^{I}\}\,E[\underline{\alpha}_i\underline{\alpha}_j]$$

$$+\,\Sigma_i\Sigma_j\Sigma_k\{u_i^{I}\}\{u_{jk}^{II}\}\,E[\underline{\alpha}_i\underline{\alpha}_j\underline{\alpha}_k] \tag{2.42}$$

An interesting alternative approach, including an extension to nonlinear problems, was provided by Liu and his coworkers. The reader is referred to Refs. [63][65][66] for more details.

Finally Shinozuka [5] has proposed that the stiffness matrix $[\underline{K}]$ be partitioned into the averaged spatial variable $[K_o]$ and the associated deviatoric part $[\underline{K}_1]$:

$$[\underline{K}] = [K_o] + [\underline{K}_1] \tag{2.43}$$

The expression

$$\{u\} = [K_o]^{-1}\,\{W\} - [K_o]^{-1}\,[K_1]\{u\} \tag{2.44}$$

gives rise to the recursive relations

$$\{u\} = [K_o]^{-1}\,\{W\}$$

$$\ldots\ldots\ldots\ldots\ldots\ldots\ldots\ldots\ldots\ldots \tag{2.45}$$

$$\{u_i\} = \{u_o\} - [K_o]^{-1}]\,[K]\{u_{i-1}\}$$

and hence

$$\{u\} = \{u_o\} - [K_o]^{-1}[K_1]\{u_o\} + [K_o]^{-1}[K_1][K_o]^{-1}[K_1]\{u\} - \dots \tag{2.46}$$

In equation (2.46) truncation before the third term characterizes the first order approximation, whilst truncation after the third term denotes the second order approximation and so on.
In the first order scheme, the corresponding expressions for equations (2.41) and (2.42) are [5] [67]

$$E[\{\underline{u}\}] = \{u_o\} \tag{2.47}$$

$$[\Sigma_{\{\underline{u}\}}] = E[[K_o]^{-1}[\underline{K}_1]^T\{u_o\}\{u_o\}^T[\underline{K}_1]^T([K_o]^{-1})^T] \tag{2.48}$$

The words "stochastic finite elements" are presently used to describe a large variety of problems of structural analysis under conditions of uncertainty. The discretization of a random field [68] is the basic feature of any stochastic finite element approach.
The stochastic finite element method is however, often used to denote simply the way for dealing with discretized structures whose properties have a stochastic nature, with external actions which can be stochastic. These techniques have the characteristic of requiring that the finite element code be manipulated, since additional computations parallel to the structural solving process are required.

When complex structural systems with spatial transient external actions (e.g., thermal and impact shocks [69]) are considered, the introduction of the appropriate modifications inside the finite element algorithm may become a formidable task.
In this case the approach proposed in Refs [70]and [6] may be more satisfactory. The method identifies a subspace of the design variable space wherein to build a polynomial approximation (response surface) of a response parameter of interest. An error term accounts for the randomness of variables, vectors and fields which are not directly represented in that subspace. A FORM/SORM method is then used to find the probabilistic properties of this response.
For reliability calculations great care must be devoted to the selection of the region of the subspace within which the response surface is to be built.

2.1.3 A RSM stochastic FEM approach

Let Y be a response variable of a structural system whose design variables X_j (j=1,...,N) are random. The response is a random function of the input variables but this function is often unknown, or complicated.
A model has been proposed by Faravelli [6][70] for expressing this dependence. A brief overview of the approach is given here.

Any vector $\{X\}_j$ expressing the spatial variability of the design variable $\underline{X}_j$ can be written in the form:

$$\{\underline{X}\}_j = X_{Aj} + \{\underline{X}'\}_j \tag{2.49}$$

where X_{Aj} is the central value of X_j and $\{\underline{X}'\}_j$ denotes deviations of $\{\underline{X}\}_j$ from X_{Aj} . The relationship between Y and the set of spatial averages $\{X_A\}$ is here analysed by the response surface method [71][72]. For this purpose, appropriate transformations y of Y and $\{x_A\}$ of $\underline{X}_A$ can be found for which a low order polynomial relationship $y(\{\underline{x}_A\})$ holds. In particular the standardized variable

$$x_{Aj} = (\underline{X}_{Aj} - E[\underline{X}_{Aj}])/(Var[X_{Aj}])^{\frac{1}{2}} \tag{2.50}$$

is introduced for any Gaussian X_{Aj}. In equation (2.50) E[.] denotes the mean value and Var [.] the variance, as usual. The transformation y of Y is selected in order to acheive greater accuracy in the response surface modelling.

The polynomial model of the dependence of y on {x}, introduced in [70], is written in matrix notation as follows:

$$Y = \theta_o + \{x_A\}^T \{\theta_1\} + \{x_A\}^T [\Theta] \{x_A\} + \epsilon_A = \hat{Y} + \epsilon_A \tag{2.51}$$

The coefficients θ_o , $\{\theta_1\}$, $[\Theta]$ are found by a regression analysis of the results obtained in numerical experiments appropriately planned [72].

The term

$$\epsilon_A = \Sigma_j (e_j + \Sigma_k e_{jk} + \ldots) + \epsilon \tag{2.52}$$

takes into account both the model error ϵ, the effects of the vectors $\{\underline{X}'\}_j$ of the deviations from the spatial averages and their interactions. The error ϵ_A can be studied by a one-way ANOVA, as shown in [70]. It can also be decomposed according to Eq.(2.52) by a multi-way ANOVA. In this case the introduction of the assumption that the effects {e} do not depend on $\{x_A\}$, makes the model more operative.

The design of the experiments necessary for developing the regression analysis, and the one-way ANOVA is discussed in detail in [6]. The model has been improved in Ref. [73] in order to compute also the correlation amongst the response variables of interest.

The validation of the model is pursued at different levels as follows[74]:

- The lack of fit contribution to the variance of the fitted response surface must be close to the variance of ϵ_A; otherwise the model must be rejected. The way of choosing one transformation y of Y, rather than another, is that the ratio $\lambda = s_1/ s_\epsilon$ is closer to 1: $s_1{}^2$ is the lack of fit sum of squares divided by the appropriate number of degrees of freedom and $s_\epsilon{}^2$ is the pure error sum of squares divided by the corresponding number of d.o.f.
- The global pure error must be almost constant over the region under

consideration; otherwise the region covered by the composite design must be reduced. This can lead the analyst to conceive a piece-wise polynomial relationship $y(\{x_A\})$ in different regions of the space of the variables x_{Aj}.

Simulation can finally be employed for producing a graphical comparison of the estimated cumulative distribution function of the response.
Suppose that the model of Eq.(2.51) has been fitted.
As the calculation of the exact cumulative distribution function (CDF) $P_y(\gamma)$ of y is difficult, several approximate procedures can be pursued. A parametric assumption about P_y can be made, for instance, and the parameters can be fitted by the method of moments. Another possibility is to use Monte-Carlo simulation, followed by a parametric or non-parametric estimation of P_y. These methods are adequate for the central part of P_y. It is possible to increase the efficiency of Monte-Carlo simulation by importance sampling techniques, but an appropriate use of Level-2 reliability methods has been shown to be very accurate.
The basic idea introduced in [75] is to evaluate $P_y(\gamma)$ as $\Phi[-\beta(\gamma)]$, where:

$$\beta(\gamma) = \pm \min\{[(\{z\} - \{\mu_{\{Z\}}\})^T [\Sigma_{\{Z\}}]^{-1}(\{Z\} - \{\mu_{\{Z\}}\})]^{\frac{1}{2}}\} \qquad (2.53)$$
$$\{z\}:y(\{z\}) = \gamma$$

where $\underline{Z}$ is the vector of all random quantities in the right-hand side of Equation (2.51) assumed with normal distribution $N(\{\mu_{\{Z\}}\},[\Sigma_{\{Z\}}])$. The procedure can be modified for non-normality (See section 2.1 and Chapter 3).

2.3 APPLICATION OF A STOCHASTIC FINITE ELEMENT METHOD

An example is taken from Ref. [74]. It is chosen for its simplicity and completeness, in addition it offers the possibility of comparing results from different approaches. In particular the approach of Ref.[6] compared with the one of Ref. [65]. For this purpose, the elastic-plastic cantilever-beam studied in [76] is used, for pursuing both the goals.

It is a cantilever-beam of length L=40" and height 10". The length-height ratio is not large enough to justify the use of beam-theory and a plane stress analysis is conducted. For this purpose the system is discretized into 16 vertical layers, each composed of 4 elements as in Fig. (2.3)
Unfortunately in Ref.[76] the data are not specified clearly. They are probably given in pounds and inches: namely, the Young modulus $E = 30.10^6$ lb/sqi; the yielding stress $\sigma_y = 25.10^3$ lb/sqi; the maximum load per unit length $w = 6.25\ 10^2$ lb/inch.
Moreover the Poisson ratio ν is probably .3 and the post-yelding modulus $E_T = 30.10^4$ lb/sqi as in Ref.[65].

The material density is given as ρ = .00776: it is probably expressed in lb.s²/(inches)^4. This means that the specific weight is 3.lb /(inches)^3 i.e. much higher than realistic values. Nevertheless, the maximum stress under dead load is 1300 lb/sqi:approximately 1/20 of the yielding stress. In the following the dead load effect is therefore neglected.

In Ref.[76] the only random field is the yielding stress $\underline{\sigma}_y$. Its mean is the nominal value, specified above, and its coefficient of variation is .1.

This variable $\underline{\sigma}_y$ is assumed to be perfectly correlated in elements belonging to the same layer. The autocorrelation function for different layers is given by:

$$R_{ij} = \exp(-|X_i - X_j| /(.5)L) \tag{2.54}$$

Also, in the present paper the post-yelding modulus E_T is assumed to be random. The same coefficient of variation and autocorrelation function of $\underline{\sigma}_y$ are assumed but the mean value is 30.10^4 lb/sqi. Both these random functions are regarded as the sum of a random variable ($\underline{\sigma}_{y1}$ and $\underline{E}_{T1}$) and a stochastic process ($\underline{\sigma}_{y2}$ (x) and $\underline{E}_{T2}$(x)):

$$\begin{aligned} \underline{\sigma}_y(x) &= \underline{\sigma}_{y1} + \underline{\sigma}_{y2}(x) \\ \underline{E}_T(x) &= \underline{E}_{T1} + \underline{E}_{T2}(x) \end{aligned} \tag{2.55}$$

In equation (2.55) the r.h.s. terms contribute equally to the variance of the l.h.s.. Moreover all the random quantities are assumed to be normally distributed. The approach proposed in this paper is used to estimate the probabilistic characteristics of the vertical displacements u(αw) of the free-end of the cantilever. (The load factor α ranges in (1, 2.25) in order to avoid large displacements effects, w being the deterministic yielding load).

The response surface model for the transformed displacement U (αw) is:

$$\begin{aligned} U(\alpha w) = \; & \theta_o(\alpha) + \theta_1(\alpha)\sigma_{y1}{}^* + \theta_2(\alpha)E_{T1}{}^* + \theta_3(\alpha)(\sigma_{y1}{}^*)^2 \\ & +\theta_4(\alpha)(E_{T1}{}^*)^2 + \theta_5(\alpha)\sigma_{y1}{}^*E_{T1}{}^* + \epsilon \end{aligned} \tag{2.56}$$

where the star (*) characterizes the standardized values of the variables. The coefficients of the model are evaluated by a regression analysis over the results of numerical experiments conducted for the structural systems defined by the coordinates of the points specified in Fig. 2.4. Each of the five central points is associated with one couple of the simulated realizations of the stochastic processes $\underline{\sigma}_{y2}$ and $\underline{E}_{T2}$ drawn in Fig. 2.5. Moreover the star points along axis σ_{y1} are associated with the first couple of realizations; the ones along axis E_{T1} are associated with the second couple of realizations. The factorial points along the positive (negative) bisector are associated with the third (the

fourth) couple of the realizations shown in Figure 2.5.
Different transformed variables U must be introduced for different values of α. The transformation and the relevant coefficients of the model equation (2.56) are summarized in Table II.

Table II. Transformed variables U and regression coefficients for some values of the load factor α. The value .0823 denotes the deterministic yielding displacement. The value of the constant A is 10000.

Transformation	θ_0	θ_1	θ_2	θ_3	θ_4	θ_5	Var[ϵ]	$s_1/s\epsilon$	α
$[A(Y-823)]^{.15}$	2.654	-0.80	-.001	.0090	-.0004	.0002	1.3 10	1.29	1.5
$[A(Y-823)]^{.95}$	1058	-278.2	-4.1	48.87	-6.54	1.09	1001	1.01	1.75
$[A(Y-823)]^{1.25}$	15682	-5503	-156	1222.7	-319.1	91.2	3.6 10^5	1.03	2.00
$[A(Y-823)]^2/A$	3290	-1753	-88.8	539.4	-61.3	49.62	36800	1.06	2.25

Fig. 2.6 provides a graphical illustration of equation (2.56), for α = 1.75. In this case the optimal transformation is nearly linear. This circumstance permits one to summarize, in a single picture, the displacement as a function of σ_{y1} and E_{T1}, and the form of the quadratic surface.
Fig. 2.6 shows the predominant role of the variable $\underline{\sigma}_{y1}$, with respect to $\underline{E}_{T1}$. This makes meaningful a comparison with the result of Ref.[76]. Firstly, a perfect agreement in the mean values is achieved. Moreover, the standard deviation, estimated from Table II , is .042" for α = 1.75: this result is also in perfect agreement with [76].

However, for α = 2.25, the standard deviation, .152", is higher than the value .12", as obtained by [76]. This confirms that, in stochastic finite element analyses, the randomness in the design variables averages may be more significant than the one in their fluctuations.

The statistical approach proposed in Ref.[70] by Faravelli therefore makes it possible:
i) to distinguish,in the variability of the response,the contribution of the randomness of the average values of the design variables from the one due to the fluctuation around them;
ii) to estimate the probabilistic cumulative distribution function of the response variable by an appropriate use of Level-2 reliability methods;
iii) to provide, in the assessment of the first two moments of the response variable, an accuracy comparable with that achieved by second-order perturbation techniques.

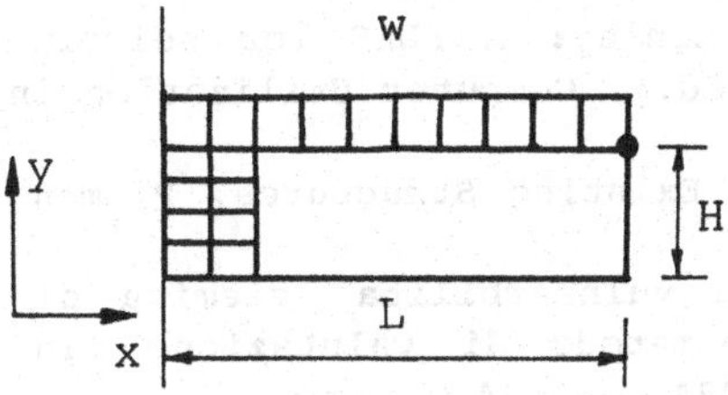

2.3

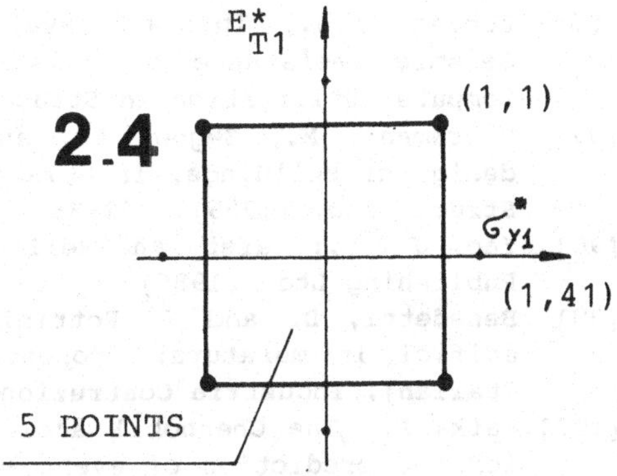

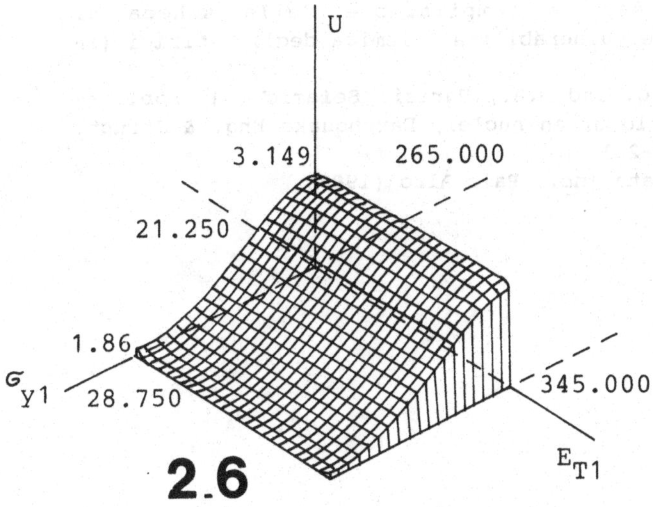

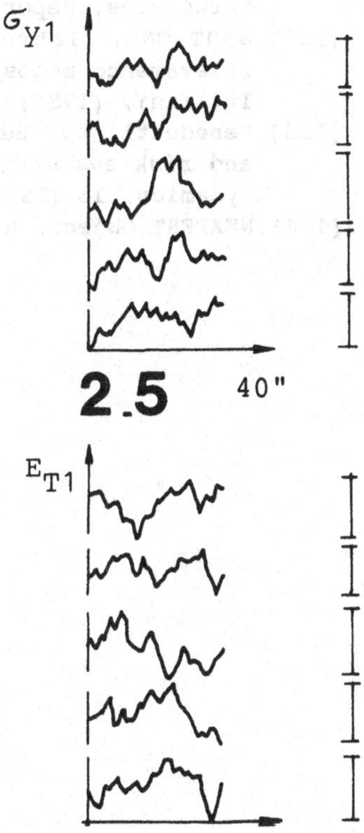

Fig. 2.3 - Geometry, loading and discretization of the beam-cantilever considered in the numerical example.

Fig. 2.4 - Experiment design in the plane of the standardized variables σ_{y1}* E_{T1}*

Fig. 2.5 - Simulated realizations of $\underline{\sigma}_{y2}$ and $\underline{E}_{T2}$.

Fig. 2.6 - Response surface expressing the displacement u(1.75 w) as a function of σ_{y1} and E_{T1}

[96] Fenves, S.J., Ibarra-Arraya, E., Bielak, J. and C.H. Thewalt: Seismic resistance of existing buildings, in Nelson J.K.(ed.), Computer Utilization in Struct. Eng. (1989), 428-458
[97] Subramani, M., Zaghw, A., and C.H., Conley: A KBES for seismic design of buildings, in Nelson J.,K. (Ed.), Computer Utilization in Struct. Eng. (1989), 342-351
[98] Yao, J.T.P.: Safety and Reliability of Existing Structures, Pitman Publishing Ltd (1985)
[99] Benedetti, D. and V. Petrini: Sulla vulnearbilita' sismica di edifici in muratura: proposta di un metodo di valutazione (in Italian). Industria Costruzioni, 18 (1984), 66-74
[100] Kafka P.: The Chernobil accident: a challenge of PSA as the tool for the prediction of event scenarios beyond the design basis and for safety improvements?, Trans. of 9th SMiRT, Vol.M (1987), 3-10
[101] Vrouwenvelder, A.: Probabilistic Model Code, Assessment of Existing Structures, Report BI-88-010, IBBC, TNO, Delft (1988)
[102] GNDT-CNR, Istruzioni per la compilazione della scheda di rilevamento esposizione e vulnerabilita' sismica degli edifici (in Italian), (1986)
[103] Benedetti, D., Benzoni, G. and M.A., Parisi: Seismic vulnerability and risk evaluation of old urban nuclei, Earthquake Eng. & Struct. Dynamics, 16 (1988), 183-201
[104] NEXPERT Object, Neuron Data Inc., Palo Alto (1987)

Two kinds of experts will be involved. One will contribute to the assessment of the fragility of his specific system component. The other will assemble the knowledge collected, for all the components, into the fragility of the whole system.
The availability of mechanical models for components and systems can make the result more objective. Otherwise each expert provides a subjective answer and all these answers are then combined into a single result. The corresponding uncertainties should also be included.

Computer science has produced recently a new research field, whose precise objective is to assign to a computer the role of an expert during consultation. The reader is referred to Ref[1] for a wider description of the nature of an expert system.
Due to the facts that: i) information in the knowledge base is incomplete and imprecise; ii) the uncertainty of information as a collection of rules and facts induces uncertainty in the validity of its conclusion, in the design of an expert system one immediately returns to the problem of managing uncertainty.
There is a number of techniques, from different disciplines, for handling uncertainty. Four general approaches are briefly reviewed here.
The "logical" model uses structured symbolic information about uncertainty and avoids numerical assessments. In particular the endorsement based approach proposed by Cohen [77] consists in the presence of reasons, called endorsments, for believing and disbelieving hypotheses. These reasons can be used to weigh evidence, to model changes in belief over inferences, to decide whether a result is certain enough to be used in subsequent reasoning, to present evidential relationships between conclusions, and to design new tasks with the goal of decreasing uncertainy.
Fuzzy reasoning is used by the "linguistic" model to quantify the extent to which the imprecise language statements match the proposition to be expressed. The term "fuzzy" was introduced by Zadeh [78] to describe sets whose membership criteria are imprecise. Uncertainty about a statement is represented by the "possibility" that is a number generated by a membership function. Zadeh has also formulated the axioms of a possibility theory [1].
The "legal" model is based on the Dempster-Shafer theory [79][80]; it relies on degrees of belief to represent uncertainty. It permits one to assign degrees to belief to subsets of hypotheses and to construct distributions over all subsets of hypotheses [1].
The "statistical enginering" model makes use of the probability calculus. Probability theory has been proved to be able to represent, fully and correctly, all reasoning under uncertainty (including vagueness) [81][82]. The Bayesian approach provides a method for updating the probability of a hypothesis given an observation of evidence. In recent years the graphical representation of probabilistic relationships between events has been developed with different terms in the literature: Bayes belief net [83], causal probabilistic network [84], causal networks [85]. A causal probabilistic network is a directed acyclic graph where the

nodes represent domain objects and the links between the nodes represent causal relations between these objects [86]. This is the path research currently under investigation, since probabilistic methods are preserved and Bayesian updating is also possible for large systems [1].
The power of the Bayesian approach is synthesized by the following example [1].
Consider a building. Without any piece of evidence, one assigns a probability.5 to the event "the building is damaged" and .5 to the event the building is not damaged". They are prior probabilities.
Then an inspection process starts. The following likelihoods are known.

Prob[no damage detected | building not damaged] = .9

Prob[no damage detected|building damaged] = .3

No damage was discovered. Then the posterior probabilities are:

Prob[building not damaged|no damage detected] =

(.9 .5) 1.0/[.9 .5 + .3 .5] = .45/.60 = .75

Prob[building damaged|no damage detected]=

(.3 .5) 1.0/.60 = .15/.60 = .25

By generalization of this simple scheme one can include new pieces of evidence in his reasoning process.

It is worth noting, however, that the first nondeterminsitic expert systems were equipped with a computational capability to analyse the transmission of uncertainty from the premises to the conclusion and associate the conclusion with: it was commonly called the "uncertainty factor" (or confidence factor). It is a numerical weight given two a fact or relationship to indicate the degree of confidence one has in it.

2.5 APPLICATION OF EXPERT SYSTEMS TO THE SAFETY ANALYSIS OF EXISTING STRUCTURES.

Several research groups of civil engineers have become active in the AI area. The reader is referred to Ref.[87] for an overview of the expert system applications in civil engineering. In particular expert system applications can be categorized into five different fields, as follows :

- Structural Engineering;
- Geothechnical Engineering;
- Constructions;
- Environmental Engineering;
- Transformation Engineering.

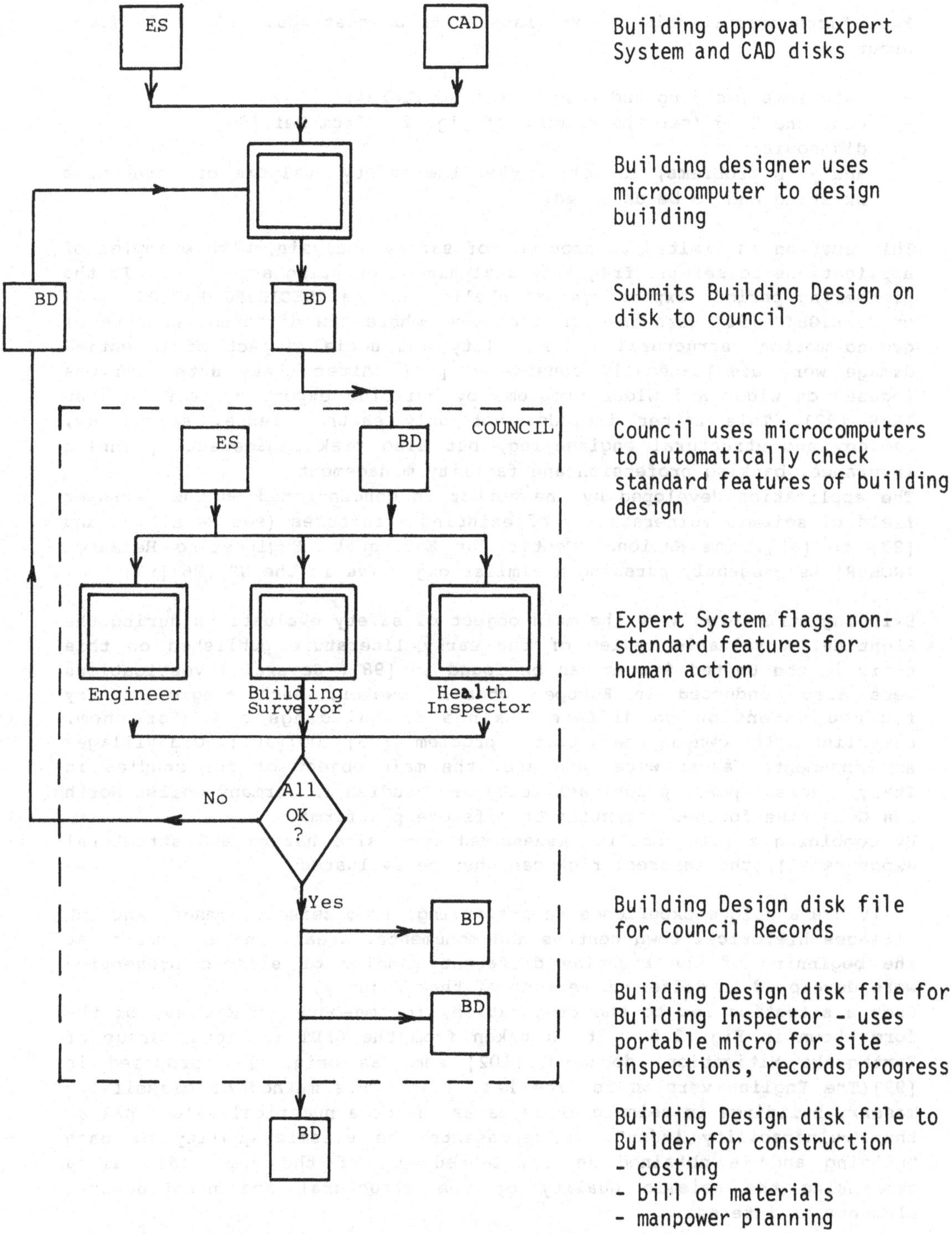

Fig. 2.7 - Scheme of expert system application to the code checking (from Ref. [88]

Four further applications area classes can be distinguished in the first group.

- materials (welding and weld defect advisors);
- code checking (see the example of fig. 2.7 from Ref.[88];
- diagnosis;
- analysis problems, as, of course, the safety analysis of structures existing and to be designed.

This section is limited to problems of safety analysis, with examples of applications to seismic fragility analysis of existing structures. In the pioneering period, expert systems shells such as "DECIDING FACTOR" [89] or "INSIGHT" [90] were used in a context where the different aspects of ground motion, structural vulnerability and social impact of potential damage were simultaneously considered [91]. After that, attention was focused on wider and wider problems by building expert systems such as IRAS [92]. This latter includes not only earth science, seismology, geology and structural engineering, but also risk management, planning insurance, banking profession and facility management.
The application developed by the author is concentrated on the narrower field of seismic vulnerability of existing structures (see Refs. [7] and [93] to [95]). The National Center for Earthquake Engineering Research (NCEER) is presently pursuing a similar objective in the US [96][97].

Existing structures were the main object of safety evaluations during the Eighties. A valuable review of the early literature published on this topic in the United States can be found in [98]. Several investigations were also conducted in Europe, where, however, each single country focused attention on different kinds of buildings and, for them, established its own approach to the problem. [99][100][101]. Old villages and monumental areas were, and are, the main object of the studies in Italy. Nuclear power plant facilities are studied in Germany whilst North Sea Countries focused attention on offshore platforms.
By combining a vulnerability assessment with site hazard and structural exposure [1], the inherent risk can then be evaluated.

Italy has a unique experience in preserving, from seismic damage, ancient villages historical town centres and monumental areas. In particular, at the beginning of the Eighties different studies of seismic prevention were developed in different regions of that country.
Such a situation led to the preparation, for masonry structures, of the form shown in Fig. 2.8. It is taken from the GNDT (National Group of Earthquake Mitigation) document [102] and was originally proposed in [99](The English version is from Ref. [1]). This method of classifying masonry buildings in seismic areas makes use of a numerical value, called the "vulnerability index". It represents the seismic quality of each building and is obtained as a weighted sum of the numerical values expressing the seismic quality of the structural and nonstructural elements of interest.

PARAMETERS	CLASS	INF. QUAL	EVALUATION ELEMENTS	EXPLANATIONS
1. Building structural organizat.	_	_	New code (A) - Strenghtened (A) - Good connec. (B) - Good joints (C) - Other (D) -	Parameter #3 Vert. str. τ_k typology t/m^2 . . . . . . Min. $A_x, A_y = A$
2. Resistant syst. qual.	_	_	see manual	Max. $A_x, A_y = B$ $\alpha = A/A_t\,;\ \gamma = B/A + 1$
3. Convent. resistance	_	-	Story number N __ Cov. area A_t (m^2) __ Area A_x (m^2) __ Area A_y (m^2) __ τ_k (t/m^2) __ Interstory h(m) __ Dead l. Y(t/m^3) __ Perm. l. Y_s (t/m^2) __	$q = \frac{(A_x + A_y)hY}{A_t} + Y_s$ $C = \frac{\alpha}{Nq} k \sqrt{(1 + \frac{2Nq}{3\alpha\gamma\tau_k})}$ $a = C/0.4$ -- Parameter #6 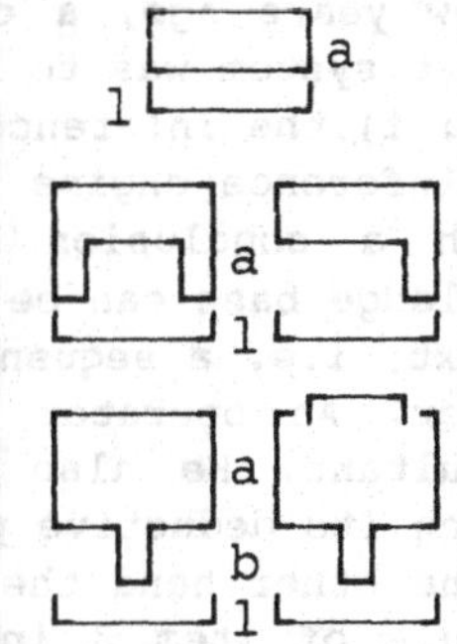
4. Building site	_	_	Land slope (%) __ Rock found. Y N Soil found. Y N Active soil Y N Max. Δh(m) __	
5. Horiz. Elements	_	_	Leapt stories Y N Elem., connect.: Rigid, Good - Flex., Good - Rigid, - Flex., - % hor.el.conn. __	$\beta_1 = a/l;\ \beta_2 = b/l$ Parameter #7
6. Plan configur.	_	_	$\beta_1 = a/l$(%) __ $\beta_2 = b/l$(%) __	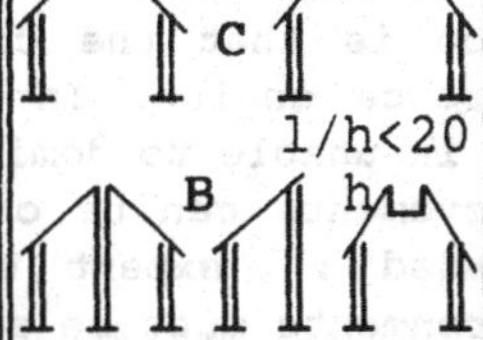
7. Vertical configur.	_	_	Δmass/mass(±%) __ T/h(%) __ Portico area(%) __ Gr.floor port. Y N	Parameter #9
8. Wall dist	_	_	Dist./thickness __	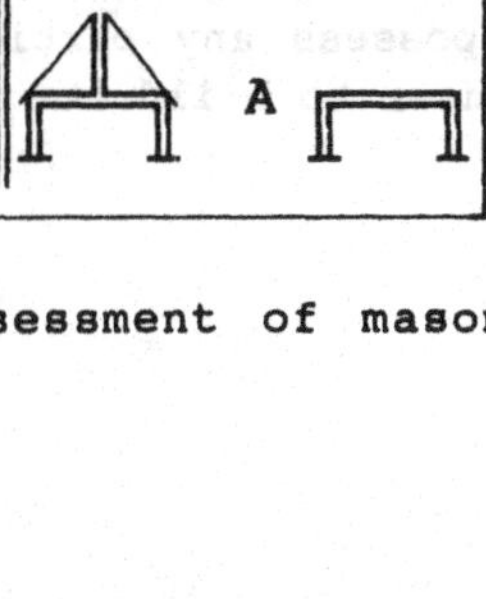
9. Roof type	_	_	Type A_ B_ C_ Ext. beams? Y N Chains? Y N Perm. load(t/m^2) __ Length (m) __ Perimeter (m) __	
10. Nostruct elements.	_	_	see manual	
11. Mainten. conditions	_	_	see manual	

Fig. 2.8 - Form for level 2 vulnerability assessment of masonry buildings, from.

One form (Figure 2.8) must be filled out for each building of interest. By means of qualitative and quantitative answers to questions on the elements, the expertise summarized in the form leads to a classification of each element into four classes (A,B,C,D). A vulnerability index is then reached by assigning, by further expertise, a weight to each answer class and a weight to each of the elements the form requires one to classify (e.g. [103]). New forms, for other kinds of buildings, which are also valuable from an architectural point of view, are presently in progress.
The use of an expert system, as a substitute for the form of Fig. 2.8, is thoroughly illustrated during the course.
The data collection cannot be automated by algorithmic computer codes since they are unable to account for descriptive (qualitative) elements. Non-algorithmic (linguistic) procedures (expert systems) are therefore required in order to implement "Level 2 Vulnerability Assistant" software.
A few years ago, a convenient approach for building a knowledge based expert system was to select a commercial "shell". It alreay includes, in fact: i) the inference enginc; 2) the text editor and 3) the consultant.
The inference engine processes the answers to some questions in order to reach a conclusion over the topic on which it was consulted. The knowledge base can be made available to the inference engine by compiling a text, i.e. a sequence of rules prepared by the expert by means of the editor. An operator, then starts the logic process by activating the consultant. He also provides the answers the inference engine needs during its deductive path (either backward or forward chaining).
On the other hand the presence of algorithmic steps (see the approximate formula of item 3 in figure 8) makes the first generation of expert systems unsatisfactory. They were not , in fact, able to alternate the qualitative and quantitative steps.
A particular shell of the second generation, also running on (portable) personal computers was used in [95]. The main advantages of this shell are as follows:
1) Local calculations are allowed.
2) External algorithmic codes can be used, without interrupting the decision process.
3) A "confidence measure" is available.
However this confidence measure is also the weak point of the shell. The reason is that the confidence calculations are driven directly by the inference engine. In other words, the expert who builds the knowledge base in unable to dominate these calculations.
Improvements can be obtained only on condition that the AI software is upgraded : expert system shells will no longer be used, but work environments must be adopted. For instance the one of Ref. [104]: it does not possess any particular way of dealing with uncertainty, but allows the user to build it.

ACKNOWLEDGEMENTS

This paper summarizes the results of researches funded by MURST (Italian Ministry of University and Research) and CNR (National Science Council), either the author or Professor L. Faravelli, of the University of Pavia, being responsible.

REFERENCES

[1] Casciati, F. and L. Faravelli: Fragility Analysis of Complex Structural Systems, Research Studies Press, Taunton, U.K. (1991)

[2] Augusti, G., Baratta A. and F. Casciati: Probabilistic Methods in Structural Engineering, Chapman & Hall, London (1984)

[3] Ditlevsen O. and O.H. Madsen: Proposal for a code from the direct use of reliability methods in structural design, JCSS, IABSE, Zurich, (1989), isbn 3-85748-059-2

[4] Casciati, F., Grigoriu, M., Der Kiureghian A. and Y.K.Wen: Report of the work group on dynamics, 1st Draft, JCSS, (1990)

[5] Shinozuka, M.: Stochastic Mechanics, Vol.I, Columbia Univ., (1987).

[6] Faravelli, L.: Finite elements analysis of stochastic nonlinear continua, in Liu W.K. and Beliytschko T. (Eds.), Computational Mechanics of Prob. and Reliab. Analysis, Elmepress, Lausanne (1989), 263-280

[7] Casciati, F. and L. Faravelli: Uncertainty treatment in a vulnerability assistant expert system, in Expert Systems in Civil Engineering, IABSE Colloquium, Bergamo (1989), 97-106

[8] Holloway, C.A.: Decision Making under Uncertainty Models and Choices, Prentice Hall, London (1979)

[9] Casciati, F. and L. Faravelli: Structural reliability and structural design optimization, Proc. ICOSSAR 85, Kyoto (1985), III, 61-70 .

[10] Rjanitsyn, A.R.: Calcul a la Rupture et Pasticite des Construction (in French), Eyrolles, Paris (1959) (translation from 1954 Russian edition)

[11] Cornell, C.A.: A probability based structural code, J. of the American Concrete Inst., 66, 12 (1969), 974-985

[12] Hasofer, A.M. and N.C. Lind: An exact and invariant first order reliability format, J.of Eng.Mech., ASCE, 100, EM1 (1974) 111-121

[13] Veneziano, D.: Contributions to Second-Moment Reliability Theory, Research Rep. R74-33, Massachussets Institute of Technology (1974)

[14] Rackwitz, R.: Reliability analysis of structural components and systems, in Thoft Christensen P.(ed.), Reliability Theory and its Application in Structural and Soil Mechanics, M.Nijhoff W. Publ. (1983), 171-214

[15] Shinozuka, M.: Basic analysis of structural safety, J. of Struct. Div., ASCE, 109(3) (1983), 721-740

[16] Schittkowskj, K.: Computational Mathematical Programming, Series F, Computer and Systems, Sciences, Springer-Verlag (1985)

[17] Schittkowskj, K.: NLPQL: a Fortran subroutine solving constrained nonlinear programming problems, Annals of Operations Research, 5, 6 (1985), 485-500

[18] Rackwitz R. and B. Fiessler: Structural reliability under combined random sequences, Computer & Structures, 9 (1978), 489-494.

[19] Breitung, K.: Asymptotic approximation for multinormal integrals, J. of Eng. Mech., ASCE, 110, 3 (1984), 357-366

[20] Tvedt, L.: On the Probability Content of a Parabolic Failure Set in a Space of Independent Standard Normally Distributed Random Variables, Section on Structural Reliability, A/S Veritas Research, Hovik, Norway (1985)

[21] Hohenbichler, M. and R. Rackwitz: Nonnormal dependent vectors in structural safety, J. of Eng. Mech., ASCE, 107, 6 (1981), 1227-1238

[22] Grigoriu, M.: Methods for approximate reliability analysis, Struct. Safety, 1 (1982), 155-165

[23] Winterstein, S. and P. Bjerager: The use of higher moments in reliability estimation, Proc. ICASP 5, Vancouver, Vol. 2 (1987), 1027-1036

[24] Ditlevsen, O., and H.O. Madsen: Probabilistic modelling of man-made load processes and their individual and combined effects, Proc. ICOSSAR '81 (1981), 103-104

[25] Tichy, M.: The science of structural action, Prof. 4th ICASP (1983), 295-321

[26] Casciati, F.: Load combination, in Lucia A.C. (ed.) Advances in Structural Reliability, D.Reidel Publ. Comp. (1987), 17-18

[27] Ang, A.H.S., and W.H. Tang: Probability Concepts in Engineering Planning and Design, John Wiley & Sons (1983).

[28] Wen, Y.K.: Structural Load Modelling & Combination from Performance & Safety Evaluation, Elsevier (1990)

[29] Larrabee, R.D. and C.A. Cornell: Combination of various load processes, J. of Struct. Div., Asce, 107(1) (1981), 223-239

[30] Grigoriu, M.: Load combination analysis by translation processes, DIALOG 6-82, Proc. Euromech 155, Lyngby (1982), 165-183

[31] Turkstra, C.J.: Theory and Structural Design Decision, Solid Mechanics Study, 2, Univ. of Waterloo (1972)

[32] Ferry-Borges, J. and M. Castanheta: Structural Safety, Lab. Nacional de Eng. Civil, Lisbon (1972)

[33] Wen, Y.K.: Statistical combination of loads, J. of Struct. Div., ASCE, 103(5) (1977), 1079-1093

[34] Casciati, F. and L. Faravelli: Load combination by partial safety factors, Nuclear Eng. and Design, 75, (1982), 439-452

[35] Faravelli, L.: A proposal on load combination for level I formats, Engineering Structures, 4 (1982), 197-206

[36] Der Kiureghian, A.: Reliability analysis, under stochastic loads, J. Struct. Div., ASCE, 106(2) (1980), 414-429

[37] Grigoriu, M.: Crossing of vector processes, in M.Grigoriu(ed) Risk, Structural Eng. and Human Error, Univ of Waterloo, Waterloo Press Waterloo, Ontario, Canada (1984), 89-112

[38] Veneziano, D., Grigoriu, M. and C.A. Cornell: Vector process models for system reliability, J. of Eng. Mech. Div., ASCE, Vol. 103 (1977), 441-460

[39] Lin, Y.K.: Probabilistic Theory of Structural Dynamics, Robert E. Krieger Publ. Co., Huntington, NY (1976)

[40] Melsa, J.L. and A.P. Sage, An Introduction of Probability and Stochastic Processes, Prentice Hall, Inc., New Jersey (1973)

[41] Roberts, J.B., and P.D. Spanos: Random Vibration and Statistical Linearization, John Wiley & Sons (1990)
[42] Grigoriu, M.: Reliability of Daniels systems subject to Gaussian load processes, Symp. on Stochastic Struct. Dyn., Univ. of Illinois at Urbana Champaign (1988)
[43] Grigoriu, M.: Reliability analysis of uncertain linear primarily secondary dynamic systems, Rep. N.89-5, School of Civil and Environmental Engineering, Cornell Univ. (1989)
[44] Wen, Y.K.: Methods of random vibration of inelastic structures, Applied Mech. Rev., 42(2) (1989), 39-52
[45] Wen, Y.K.: Equivalent linearization for hysteretic systems under random excitation, J. of Appl. Mech., Vol.47 (1) (1980), 150-154
[46] Spanos, P.D.: Stochastic linearization in structural dynamics, Applied Mechanics Rev., 34(1) (1981), 1-6
[47] Baber, T.T. and M.N. Noori: Random vibration of pinching, hysteretic systems, J. of Eng.Mech. Div., ASCE, 110(7) (1984), 1036-1049
[48] Casciati, F. and L. Faravelli: Reliability assessment of nonlinear random frames, Nuclear Eng. & Design, 90 (1985), 341-356
[49] Park, Y.J., Wen Y.K. and A.H.S., Ang: Two-dimensional random vibration of hysteretic structures, J. of Earthquake Eng. & Struct. Dyn., 14 (1989), 543-557
[50] Casciati, F. and L.Faravelli: Stochastic equivalent linearization for 3-D frames, J. of Eng. Mech., ASCE, Vol. 114, (10) (1988), 1760-1771
[51] Casciati, F. and L.Faravelli: Hysteretic 3-dimensional frames under stochastic excitation, Res Mechanica, 26 (1989), 193-213
[52] Casciati, F.: Stochastic dynamics of hysteretic media, Structural Safety, 6 (1989), 259-269
[53] Wen, Y.K. and H.-C.Chen: On fast integration for time-variant structural reliability, Prob. Eng. Mech., 3 (1987), 156-162
[54] Cambou, D.: Applications of first-order uncertainty analysis in the finite element method in linear elasticity, Proc. ICASP 2, Aachen, (1975), 67-87
[55] Esteva, L.: Second-moment probabilistic analysis of statistically loaded non-linear Structures, Proc. ICASP 2, Aachen (1975), 115-130
[56] Handa, K. and K. Anderson: Application of finite element methods in the statistical analysis of structures, Proc. ICOSSAR 81, Trondheim (1981), 409-417
[57] Hisada T. and S. Nakagiri: Stochastic finite element method developed for structural safety and reliability, Proc. ICOSSAR 81, Trondheim (1981), 395-408
[58] Lawrence, M.: Basis random variables in finite element analysis, Int. J. of Num. Meth. in Eng., 24 (1987), 1849-1863
[59] Righetti, G. and K. Harrop-Williams: Finite element analysis of random soil media, J. of Geotechnical Eng., 114(1) (1988), 59-75
[60] Der Kiureghian A. and L.P. Liu: Finite element reliability methods, Lecture Notes for Structural Reliability: Methods & Applications, Berkeley (1989)

[61] Der Kiureghian A.: Numerical methods in structural reliability, Proc. 4th ICASP, Florence (1985), 769-784
[62] Nakagiri, S.: Stochastic Finite Element Method: An Introduction, (in Japanese), Baifiukan (1985)
[63] Liu, W.K., et al.: Transient probabilistic systems, Computer Methods in Applied Mechanics and Engineering, 67 (1988), 27-54
[64] Nakagiri, S.: Fluctuation of structural response, why and how, JSME Int. J., 30(261) (1987), 369-374
[65] Liu, W.K., Belytschko T. and A. Mani: Random field finite elements, Int. J. for Num. Meth. in Eng., 23 (1986), 1831-1845
[66] Liu, W.K. and T., Belitschko (eds.): Computational Mechanics of Proabilistic and Reliability Analysis, Elmepress Int., Lausanne (1989)
[67] Szentivanyi, B.: Computer solution of stochastically linear bar structure, a discrete time simulation approach, Zeszyty Naukowe Politechniki Poznanskiej, 26 (1981), 107-108
[68] Vanmarke E. et al.: Random fields and stochastic finite elements, Structural Safety, 3 (1986), 143-166
[69] Faravelli, L. and D., Bigi: Stochastic finite elements for crash problems, Struct. Safety, 8 (1990), 113-130
[70] Faravelli, L.: Response-surface approach for reliability analysis, J. of Eng.Mech., 115, 12 (1989), 2763-2781
[71] Olivi, L.: Response surface methodology in risk analysis, in Synthesis and Analysis Methods for Safety and Reliability Studies, Plenum Press (1980)
[72] Petersen, R.G.: Design and Analysis of Experiments, M.Decker Inc, New York (1985)
[73] Faravelli, L.: Response variables correlation in stocahstic finite element analysis, Meccanica, 22, 2 (1988), 102-106
[74] Faravelli, L.: Stochastic finite elements by response surface techniques, in Computational Probabilistic Methods, ASME-AMD, Vol.93 (1988), 197-203
[75] Veneziano, D., Casciati, F. and L. Faravelli: Method of seismic fragility of complicated systems, 2nd CSNI Meeting on Prob. Meth. in Seismic Risk Assessment for Nuclear Power Plants, Livermore (1983), 67-88
[76] Liu, W.K. et al : Applications of probabilistic finite element methods in elastic plastic dynamics, J. of Eng. For Industry (ASME), 109 (1987), 1-8
[77] Cohen, P.R.: Heuristic Reasoning about Uncertainty: an Artificial Intelligence Approach, Pitmans, Boston (1985)
[78] Zadeh, L.A.: The role of fuzzy logic in the management of uncertainty in expert Systems, Fuzzy Sets and Systems, 11 (1968), 199-228
[79] Dempster, A.P.: A generalization of bayesian inference, J. of the Royal Statistical Society B, 30 (1968), 205-247
[80] Shafer, G.: A Mathematical Theory of Evidence, Princeton Univ. Press, Princeton (1986)

[81] Lindley, D.V.: The probabilistic approach to the treatment of uncertainty in artificial intelligence and expert systems, Stat. Science, 3 (1987), 17-24
[82] Spiegelhalter, D.J.: Probabilistic reasoning in predictive expert systems, in Kanal L.N. and Lemmer J.F. (eds.), Uncertainty in Artificial Intelligence Elsevier Sc. Publ. (1986), 47-68
[83] Pearl, J.: Fusion propagation and structuring in belief networks, Artificial Intelligence, 28 (1986), 9-15
[84] Andreassen S., Woldbye M., Falck., B. and S.K. Andersen, MUNIN - A causal probabilistic network for interpretation of electromyographic findings, Proc. 10th Int. Joint Conf. on Artificial Intelligence, Milan (1987), 366-372
[85] Lauritzen, S.L. and D.J. Spiegelhalter: Local computations with probabilities on graphical structures and their application to expert systems, J. Royal Statistical Society B, 50(2) (1988), 157-224
[86] Gherardini, P.: Implementazione object-oriented di un modello di calcolo per reti causali (in Italian), Proc. XXXV Sc. Meeting of Italian Statistical Society, Padova (1990)
[87] Maher, M.L., (ed.): Expert Systems for Civil Engineers: Technology and Application, American Society of Civil Engineering, ASCE, (1987)
[88] The Building Surveyor (Off. J. of The Australian Institute of Building Surveyors), 7(8) (1988)
[89] Campbell, A. and S. Fitzgeral: The Deciding Factor, User's Manual, Software Publ. Co. (1985)
[90] INSIGHT, Knowledge System, Level Five Research, Merbourne Beach, Florida (1985)
[91] Miyasato, G., Dong, W.M., Levitt, R.E., Boissonade A.C. and H.C. Shah: Seismic risk analysis system, Proc. Symp. Expert Systems in Civil Engineering, Seattle (1986) 121-132
[92] Dong, W., Wong, F., Chiang, W., Kim, J.U. and H.C. Shah: An integrated system for seismic vulnerability and risk for engineering facilities, in Nelson J.K.(ed) Computer Utilization in Structural Engineering, ASCE (1989), 408-417
[93] Casciati, F. and L. Faravelli: L'impiego di sistemi esperti in ingegneria sismica (in Italian), Proc. 3rd Conf. Earthquake Eng. in Italy, Rome (1987), 199-210
[94] Casciati, F. and L. Faravelli: Individuazione di problemi di meccanica dei solidi suscettibili di inquadramento in sistemi esperti, (in Italian), Proc. 9th Conf. AIMETA (Ital Assoc. of Theoretical and Applied Mechanics), Bari (1988), 553-556
[95] Casciati, F. and L., Faravelli, Seismic vulnerability via knowledge based expert systems, Brebbia C.A. (ed.), Structural Repair and Maintenance of Historical Buildings, Computational Mechanics Publ., Southampton (1989), 299-307

The whole procedure becomes operative at a practical level when it is associated with appropriate software. This includes the pre- and post-processor for any input-output finite element black-box. The scenario in which the pre-processor operates assumes that it is possible to identify:
a) some variables such that the response surface expresses the functional dependence of the response variable of interest on them;
b) a group of remaining variables (to be grouped in a single vector) whose contribution to the response variable is confused with the others and is mixed with the contributions of the elements listed in the next point c);
c) some stochastic processes in time and some random fields whose contribution to the response variables is confused in a single error term ϵ.
The scenario in which the post-processor is operative considers some response variables for which the marginal distribution is computed. For each couple of variables, the joint distribution can also be evaluated.
The previous features were implemented in the computer code RESFEM (Response surface Stochastic Finite Element), capable of running on personal computers. The use of this code is thoroughly illustrated, by practical applications, during the course.

2.4 MANAGING UNCERTAINTY IN EXPERT SYSTEMS

It has been already observed in Section 2.1.3 that technical problems with predominant theoretical difficulties give rise to parallel levels of approach to the problem. All these levels will make intensive use of mathematical models. In this context the applied mechanics nature of the structural engineer leads him to select a way of managing uncertainty with algebra as coherent as possible with the one governing the mechanical models. The axiomatic nature of the Theory of Probability fits this coherency requirement perfectly.

However, on the other side, at the engineering level, a decision must be taken with incomplete information and mechanical models that cannot always be very accurate. One cannot wait for the collection of adequate statistics, or the calibration of the mathematical models. From this point of view (currently prevailing when detailing existing plants and/or structures) probabilistic methods can be applied only if the "way of selecting numbers" is clearly explained.

Moreover, when dealing with structural systems one meets both uncertainty and randomness. Probabilistic models for both of them, are often unavailable; this is mainly true when dealing with large structural systems. In this case one generally gathers a pool of experts who contribute in their own language to system fragility evaluation. Of course, fragility is no longer the probability of failure of the system versus the loading intensity. It is just the confidence regarding the satisfactory behaviour of the system.

Chapter 3

METHODS FOR STRUCTURAL RELIABILITY COMPUTATIONS

P. Bjerager
Veritas Sesam Systems A.S., Hovik, Norway

SUMMARY

Probability computation methods for structural and mechanical reliability analysis are presented. Methods for random variable, random process and random field reliability models are included, with special emphasis on random variable models. Recent developments are described, assuming the reader to be familiar with earlier methods. The presently available state-of-the-art computation methods are evaluated, and their merits are discussed and compared.

3.1 INTRODUCTION

3.1.1 Structural reliability theory

Designing and operating engineering systems involves decision making under uncertainty. This uncertainty may relate to both the necessary capacity and the actual capacity of a system. The uncertainty is due to random fluctuations of significant physical quantities, due to limited information on these physical quantities, and due to model idealizations of unknown credibility introduced because of lack of knowledge as well as a need for operability of the engineering model.

Within the field of civil engineering the solution of such decision making problems have been facilitated by the use of codes of practice. In the simplest form these codes simply record common practice developed over a long term experience of successful design. However, shortcomings of such codified experience are evident when a problem arises outside the area on which the experience is based. Moreover, to establish consistent codes a theoretical quantification of the uncertainties present is necessary. Such theoretical considerations may, furthermore, lead to engineering systems which are optimal in a well defined sense.

Structural reliability theory is a discipline for dealing with uncertainties in a consistent and rational way. The theory has developed rapidly in the last two decades, up to the present stage of providing a conceptually and operationally satisfactory reliability methodology. It allows for a rational comparison between alternative structural designs, maintenance plans, etc., and it is based on simple principles of including both physical, statistical and model uncertainty. The representation of uncertainty may be based on recorded physical observations as well as subjective, professional judgments. As more information becomes available it can be included in the reliability model by using Bayesian updating methods.

An essential result of a structural reliability analysis is the *reliability measure*. The measure may depend on a number of assumptions adopted by the analyst. The reliability measure shall be interpreted as an *engineering factor* expressing the current knowledge/information about the structure and its environment as regards the structure's ability to meet the considered requirements - under the assumptions set forth in the analysis (for example assumptions regarding probability distribution types and the structural models adopted). The reliability measure introduces thereby an ordering of the considered structures with respect to reliability that can be used in comparative studies.

The reliability measure is often taken as the *reliability index*, in general defined as a monotone function of the *failure probability*. There are several reasons for using the reliability index (instead of probability directly); the reliability index has a historical significance, it has a clear geometrical interpretation in a number of basic cases, it is an essential component in first order reliability methods, it has a certain simple relation to common partial safety factors, and the use of it emphasizes its role as an engineering factor.

Structural reliability concepts may be used directly in the design and operation phases of a structure. Another important application so far is its use in the formulation and rationalization of structural codes of practice. This has resulted in significant changes in code formats and has lead to consistent calibration and optimization of the adjoined safety factors. The practical use of structural reliability methods is now possible, due to the availability of a number of automated reliability computation methods.

3.1.2 Reliability computation methods

The theory and methods of structural reliability have developed significantly during the last 10-15 years. In this period research on both philosophical and conceptual issues, as well as on reliability and sensitivity computation methods has taken place. The field has now reached a stage where use of the developed methodology is becoming widespread. Besides the constant use of probabilistic methods in the calibration of technical standards and codes, the methods find use in, for example, the design and operation of unique structures such as offshore and bridge structures, where experience from similar structures may not exist.

A prerequisite for the general use of probabilistic methods in everyday design is the availability of automatic and efficient reliability and sensitivity computation methods. Even though the computation methods - at least as far as the random variable models are concerned - must be considered to be very well developed by now, there is still a need to strike a balance between the structural and mechanical modeling and the computational ability. To understand this balance it is important to know the characteristics of the present probability computation methods. It is the aim of the present paper to convey to the reader such an understanding.

Even though the balance between computational ability and mechanical realism in reliability models has at times, lead to idealized models that are met by some scepticism or criticism among engineers familiar with (advanced) mechanical modeling, there is now a development towards the use of probabilistic models within general purpose structural analysis methods such as the finite element method, see for example [Vanmarcke et al, 1986; Der Kiureghian and Ke, 1987; Shinozuka, 1987; Liu et al, 1987 and 1988; Liu and Der Kiureghian, 1988a]. This development is important since it enables standard mechanical models to be used in probabilistic analysis. Furthermore, it puts the probability computation methods to a test, which may trigger new and more general methodologies.

The methods and application of structural and mechanical reliability theory have been documented in an increasing number of text books such as [Ditlevsen, 1981; Ang and Tang, 1984; Augusti et al, 1984; Madsen et al, 1986; Melchers, 1987]. The present paper gives an outline of the most recent developments of the computation methods assuming the reader to be familiar with reliability theory. Reference is made primarily to the latest papers, leaving references to subjects published earlier to be found in the books above. The probability computation methods are evaluated with respect to limitations, and further research needs are identified.

3.2 CLASSES OF RELIABILITY MODELS

Mathematical reliability models can be classified according to the level of description applied in the uncertainty modeling. In this section, models are grouped into 1) random variable models, 2) random process models, and 3) random field models. For each type of model computation methods are discussed. Emphasis is given to the simplest type of problems, the random variable models, for which general and practical computation methods exist. The two most important classes of computation method - first and second order reliability methods and Monte Carlo simulation - are discussed separately in succeeding sections.

3.2.1 Random variable reliability models

A random variable reliability problem is defined by the safety margin $G(\mathbf{X})$, where G is the failure function, and $\mathbf{X}$ is a set of n random physical basic variables. Failure is defined by the event $\{G(\mathbf{X})\le 0\}$, and $\{G(\mathbf{X})>0\}$ identifies a safe state. The failure probability is

$$p = P[G(\mathbf{X})\le 0] = \int_{G(\mathbf{x})\le 0} f_{\mathbf{X}}(\mathbf{x})d\mathbf{x} \tag{3.1}$$

where $f_{\mathbf{X}}(\mathbf{x})$ is the multivariate density function of $\mathbf{X}$. If some of the basic random variables are discrete, the integration over the corresponding densities is substituted by a summation over finite probabilities. Depending on the application, the function G may also be referred to as the limit state function, the performance function, the state transition function, the event function, or, simply as the G-function.

The considerations above are based on the assumption that the possible structural performances have been divided into failed states $\{G(\mathbf{x})\le 0\}$ and safe states $\{G(\mathbf{x})>0\}$. However, it is noted that in general it may not be possible to set up a sharp discrimination between these two structural states. Instead a utility function $U(\mathbf{x})$ may be defined for each outcome of the basic variables $\mathbf{X}$. The structure can then be assessed in terms of the expected utility given as

$$E[U]=\int_{R^n} U(\mathbf{x}) f_{\mathbf{X}}(\mathbf{x}) d\mathbf{x} \tag{3.2}$$

It is seen that if the utility function $U(\mathbf{x})$ is taken as one in the safe set and zero in the failure set, the expected utility equals the reliability, i.e.

$$E[U]=1-p \tag{3.3}$$

The primary purpose of reliability computation methods for random variable reliability models is to evaluate the multi-dimensional integral in Eq. 3.1. Only a few analytical and exact results are known. However, these results are extremely important in analytical computation methods, and these results are outlined in Sec. 3.4 below. Standard numerical integration techniques are generally not feasible for high-dimensional problems (large n), and in general, either Monte Carlo simulation (MCS), or analytically based first- and second-order reliability methods (FORM/SORM) must be used, see [Rubinstein, 1981; Augusti et al, 1984; Madsen et al, 1986]. Computation methods for random variable reliability problems are extensively developed and can handle many practical and realistic problems. Commercial software for implementing the computation methods is available, for example the probabilistic reliability and sensitivity analysis program [PROBAN].

The failure function G is often specified in a "structured" way, i.e., the function is constructed following basic concepts of *classical reliability theory*, e.g. [Barlow and Proschan, 1981]. Thus, a failure function is specified for a set of binary-states components, which can be arranged into systems. For example, the state of the ith component is described by the failure function $G_i(\mathbf{x})$, such that failure is defined by $\{G_i(\mathbf{x})\leq 0\}$, and $\{G_i(\mathbf{x})>0\}$ identifies a safe state of the component.

Basic *system representations* are shown in Fig. 3.1, i.e., series systems, parallel systems, series systems of parallel sub-systems, and parallel systems of series sub-systems. By the two latter representations, a system can be described by minimal cut sets and minimal path sets. Representations in terms of fault trees, event trees and general reliability networks can therefore be described by the basic system types above, and it is possible to have automatic conversion between such representations.

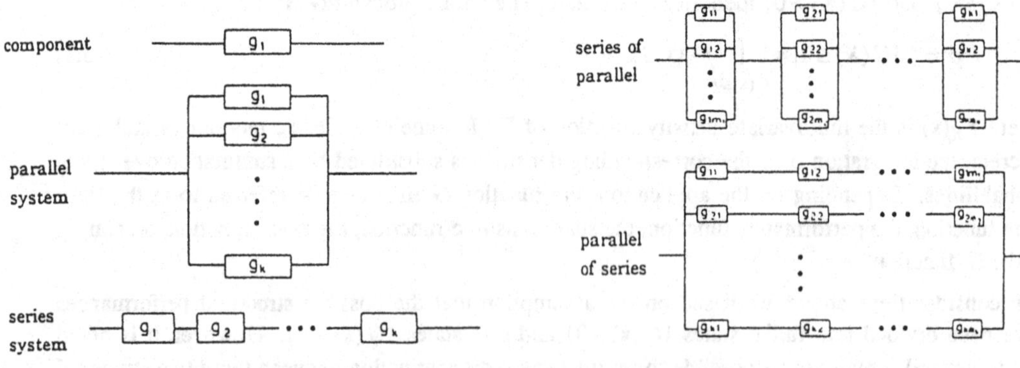

Fig. 3.1. Illustration of basic systems representations. It is noted that the lower case notation g for the failure functions is used in the transformed (standard normal) space - see below.

Considering the two simplest systems, the safety margins for a series system of m components is

$$G(\mathbf{X}) = \min_{i}^{m} \{G_i(\mathbf{X})\} \tag{3.4}$$

and that for a parallel system of m components is

$$G(\mathbf{X}) = \max_{i}^{m} \{G_i(\mathbf{X})\} \tag{3.5}$$

The failure function for the higher level systems is constructed equivalently. It is noted that each component is not only represented by an indicator variable, as in classical reliability theory, but by a random vector reliability model expressed in terms of a failure function that can reflect an underlying physical model. Therefore, within the common model framework in structural and mechanical reliability theory, it is possible to model very general stochastic dependencies in the system, for example by having basic variables common to the components, or by using a multivariate distribution with stochastic dependence between the basic variables. Typically, the function would be based on an underlying mechanical model, and the function could be defined in terms of a separate computer program, such as a finite element code. The numerical value of the failure function may have a relevant physical interpretation, but for many problems only the sign has importance. In that case, a negative sign designates a failure state, and a positive sign a safe state.

The two broad classes of probability computation methods, MCS and FORM/SORM, are often presented as being *competing* probability integration methods. However, it seems much more fruitful - and correct - to consider them *complementary*, not least for the simple reason that it is easy to identify problems where MCS is preferable to FORM/SORM, and vice versa. For example, since FORM/SORM are analytical probability integration methods, they do not apply when the problem at hand does not fulfill - or has not been modeled to fulfill - the necessary analytical requirements. In the case where the basic variables are discrete, the necessary transformation between the physical x-space and the standard normal u-space does not exist. On the other hand, for the large class of engineering problems where FORM/SORM methods *do* apply, and p is very small ($10^{-4} - 10^{-8}$), FORM/SORM is generally preferable to MCS.

The development of MCS and FORM/SORM can in fact be considered to be two parallel processes with mutual benefits - and common problems. One of the major breakthroughs within FORM/SORM was the identification of a generally applicable method for the transformation of a random vector $\mathbf{X}$ with a continuous distribution function $F_{\mathbf{X}}(\mathbf{x})$ into a standard Gaussian vector $\mathbf{U}$ (zero means, unit variance and independent components) [Hohenbichler and Rackwitz, 1981]. In simulation, a quite similar transformation is used when outcomes of the random vector are generated by the inverse transformation method, see [Rubinstein, 1981]. In this case, $\mathbf{X}$ is transformed into a set of uniformly distributed variables.

In order to increase the efficiency of MCS, importance sampling has been suggested, and tested, see e.g. [Rubinstein, 1981; Harbitz, 1983; Shinozuka, 1983; Augusti et al, 1984; Melchers, 1987]. The sampling density is typically located in a region in the failure set $\{\mathbf{x} \mid G(\mathbf{x}) \leq 0\}$ of relatively high density, and a common suggestion has been to centre the sampling distribution in the neighbourhood of the most likely failure point. This point - here defined as the point of maximum density on the failure surface in u-space - is of unique importance in FORM, and MCS, with such an importance sampling, therefore has strong similarities to a FORM-computation. In fact, often

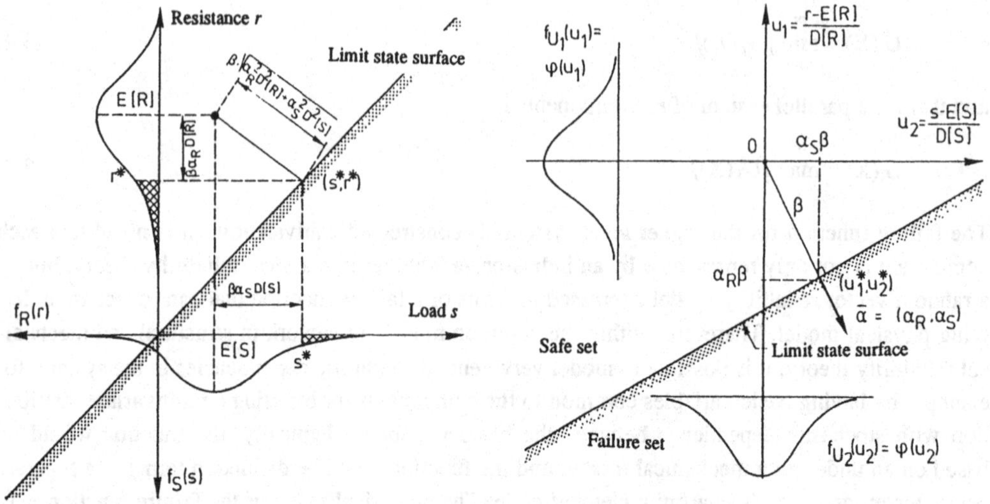

Fig. 3.5. Geometrical interpretation of elementary reliability index. Left: In (r,s)-space. Right: In u-space.

The elementary reliability index as defined by Eq. 3.9 turns out not to be an appropriate reliability measure in this case (and the more general case in the next section), because it depends on the formulation of the G-function which is not unique for a given reliability problem. In order words, equally valid and equivalent models may yield two different values of the elementary reliability index. This is known as the invariance problem of the elementary reliability index [Ditlevsen, 1973].

Instead, the reliability index must now be computed from the generalized definition based on the failure probability, i.e., as

$$\beta=\Phi^{-1}(1-p) \tag{3.16}$$

where the failure probability p is computed by some method (see later). The reliability index has then, in general, no geometrical interpretation. However, if the reliability index is approximately computed by a first order reliability method, this first order reliability index is the minimum distance from the origin in u-space to the (curved) limit state surface. The first order reliability index is also known as the Hasofer-Lind reliability index or the *geometric reliability index*, Fig 3.6. Both the elementary reliability index and the geometric reliability index originated as concepts in second moment structural reliability theory [Cornell, 1969; Hasofer and Lind, 1974; Ditlevsen, 1981].

3.3.3 R(X) and S(X) nonlinear functions

If R and S are functions of a set of basic random variables **X**, i.e. $R=R(\mathbf{X})$ and $S=S(\mathbf{X})$, the safety margin $G(R,S)$ turns into a general nonlinear safety margin of the form $G(\mathbf{X})$ treated in the subsequent sections.

Hagen, 1991; Hagen and Tvedt, 1991]. The outcrossing is formulated as a zero down-crossing of a differentiable scalar process, and the mean crossing rate is obtained as a sensitivity measure of the probability for an associated parallel system reliability problem [Madsen, 1990]. The vector process may be Gaussian, non-Gaussian, stationary or non-stationary, and the limit state surface may be time-dependent. By the same methodology, n'th order joint crossing rates and n'th order joint distributions of local extremes can be computed for the same classes of vector processes [Hagen, 1991].

For crossings of surfaces defined by the random vector **X**, the outcrossing can be considered conditionally on the outcomes of these variables. The failure probability is computed by a nested application of FORM/SORM, [Madsen and Zadeh, 1987; Wen, 1987; Madsen and Tvedt, 1988]. For time-dependent failure surfaces, and/or non-stationary processes the outcrossing rate becomes time-dependent. For history-dependent problems, no analytical solution method is known.

Compared to random vector reliability problems, the analytical computation methods for random process models are rather restricted. The FORM/SORM for Gaussian processes are not generally applicable due to the lack of a simple transformation of continuous non-Gaussian processes into Gaussian processes. For superimposed processes of significant different time scales - such as for example mean wind fluctuations and wind turbulence fluctuations - an accurate result may be obtained by discretizing the slow process into square wave processes of constant pulse duration [Bjerager et al, 1988b]. Further development of the analytical computation methods for random process models is to be expected.

3.2.3 Random field reliability models

A random field reliability problem is described by a function $G(\mathbf{X},\mathbf{Y}(t),\mathbf{Z}(\mathbf{q},t))$, where **X** is a set of random variables, $\mathbf{Y}(t)$ is a set of random (vector) processes, $\mathbf{Z}(\mathbf{q},t)$ is a set of random (vector) fields, t is time, and **q** is, for example, the spatial coordinates.

Random fields are ideal for modeling spatial variations such as those in material properties, or the random variation over the structure of the distributed load, for example the wave loading on an offshore structure. A general treatment of random fields for engineering applications is given by [Vanmarcke, 1984]. The present computational abilities for random field reliability problems are very limited. A recent review on extremes of fields is given by [Faber and Rackwitz, 1988]. The analysis of random fields leads to multi-dimensional integrals with some similarity to the ones known from random variable problems [Rackwitz, 1988].

So far, random fields have mostly been applied for modeling purposes, for example in the underlying stochastic modeling for probabilistic finite element models, as in [Liu et al, 1986], for the element properties as well as for the nodal forces. Because of the discretization, the resulting reliability model becomes of the random variable type, and the *computation* methods to be used will be those for random variable reliability models.

3.3 THE R - S RANDOM VARIABLE RELIABILITY MODEL

An elementary random variable reliability model that is used in many situations is the simple $R-S$ model in which a resistance R (or strength or capacity) of a system is compared to the loading S on (or requirement to) the system. Assuming both quantities to be random, the safety factor

$$K = \frac{R}{S} \tag{3.6}$$

is also random. Whether the system behaves safely $\{K > 1\}$ must therefore be treated probabilistically, i.e. what is the *probability* that the event $\{K > 1\}$ occurs.

A simple physical system that can be represented by the R–S model is the bar shown in Fig. 3.2, where also the distribution of the resistance R and of the load effect S are shown. It is not uncommon to show these two distributions in a one-dimensional diagram as in Fig. 3.2. However, as will become evident in the following, it is useful and more appropriate to consider the distributions in a two-dimensional diagram.

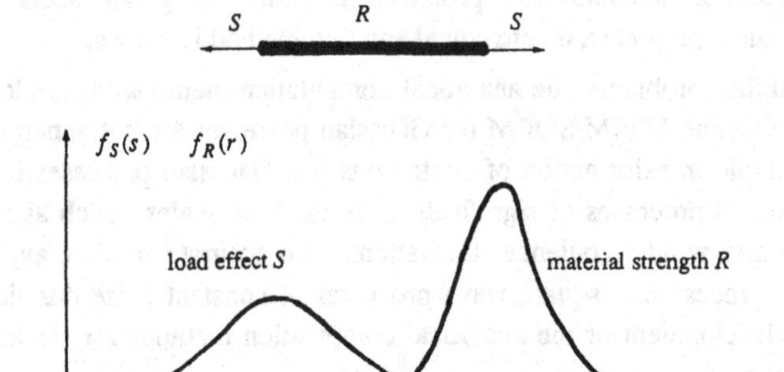

Fig. 3.2. The basic R–S reliability model - showing R and S on the same axis.

R and S can be modeled as functions of a number of factors. If the bar belongs to a larger structure, the distribution of the load effect and of the resistance will vary from bar to bar. For example, the resistance could be expressed as a product $R = R_1 R_2 R_3$, where R_1 represents the physical uncertainty in the material property, R_2 the level of on-site quality control and R_3 the scatter representing the number of plants from which the material originates. Correspondingly, the load effect may be factorized into $S = S_1 S_2 S_3$ where S_1 represents the physical uncertainty of the environmental loading, S_2 represents model uncertainty in the global stress analysis of the system, and S_3 represents uncertainty in the local stress analysis of the sectional stresses. Adopting such idealized modeling, the R–S model has been applied extensively to understand important aspects of the influence of different uncertainty sources.

The safety margin for the R–S reliability model is

$$M = G(R, S) = R - S \tag{3.7}$$

and the limit state surface $G(r, s) = 0$ is shown in Fig. 3.3 together with the marginal and joint distribution of (R, S). The safety margin is a random variable and the failure probability is $P[M < 0]$, Fig. 3.4.

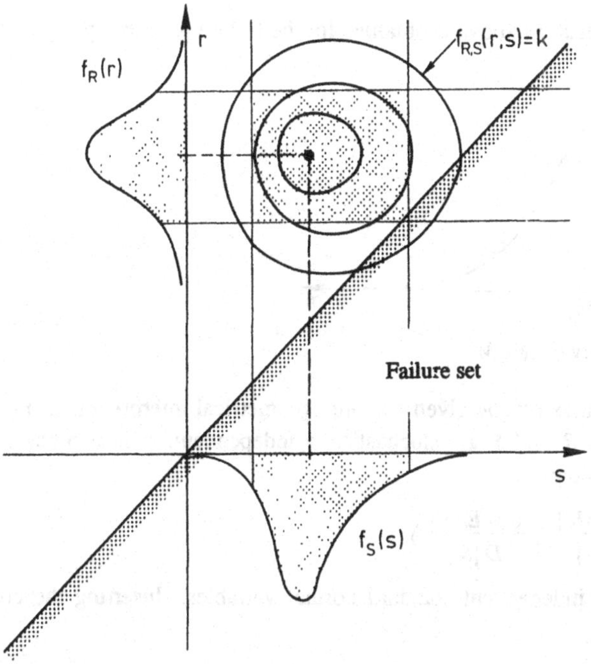

Fig. 3.3. The $R-S$ model in a two-dimensional illustration showing the marginal distributions $f_R(r)$ and $f_S(s)$, the joint distribution $f_{R,S}(r,s)$, and the limit state surface $G(r,s)=0$.

3.3.1 R and S normally distributed

If R and S are assumed to be normally distributed, the safety margin M is also normally distributed, and the failure probability can be easily computed as

$$p = P[M \leq 0] = P[\frac{M - E[M]}{D[M]} \leq -\frac{E[M]}{D[M]}] = P[U \leq -\beta] = \Phi(-\beta) \tag{3.8}$$

where U is a standard normal variable, Φ is the corresponding probability distribution function, $E[]$ denotes the expectation (mean value)a and $D[]$ denotes the standard deviation. β defined by

$$\beta = \frac{E[M]}{D[M]} \tag{3.9}$$

is known as the *elementary reliability index*. Considering the safety margin in Fig. 3.4, β can be interpreted as the number of standard deviations the failure event $M=0$ is away from the mean value $E[M]$.

Instead of utilizing the distribution of M (which in this Gaussian case was known and easily computed), the failure probability can be computed by direct integration, i.e.

$$p = \int_{-\infty}^{\infty} \int_{r=-\infty}^{s} f_{R,S}(r,s) dr ds = \int_{-\infty}^{\infty} \int_{r=-\infty}^{s} f_R(r) f_S(s) dr ds = \int_{-\infty}^{\infty} F_R(s) f_S(s) ds \tag{3.10}$$

It is assumed in the second term of Eq. 3.10 that R and S are stochastically independent, whereby the joint distribution can be expressed as the product of the marginal distributions. Inserting the normal densities, the same result as above is obtained for the failure probability.

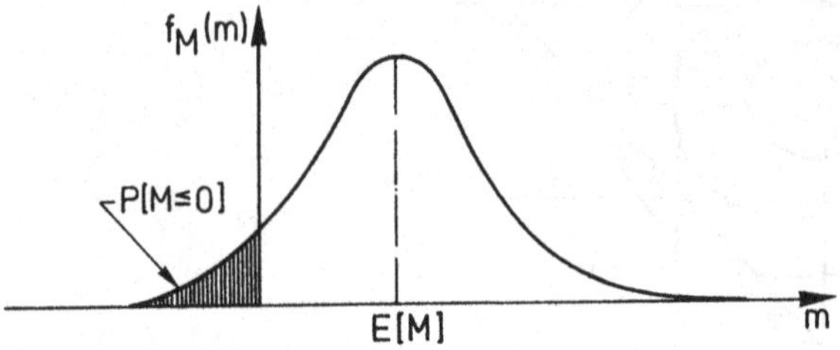

Fig. 3.4. Illustration of the safety margin M.

The elementary reliability index can be given a simple geometrical interpretation. For the sake of simplicity it is assumed that R and S are stochastically independent. The two basic variables R and S are standardized as follows

$$(U_R, U_S) = \left(\frac{R - E[R]}{D[R]}, \frac{S - E[S]}{D[S]}\right) \tag{3.11}$$

where U_R and U_S are two independent standard normal variables. Inserting the corresponding expressions for R and S

$$(R, S) = (E[R] + U_R D[R], E[S] + U_S D[S]) \tag{3.12}$$

into the expression for the safety margin, we obtain

$$M = E[R] + U_R D[R] - E[S] - U_S D[S] \quad \rightarrow$$

$$\bar{M} = \frac{E[R] - E[S]}{\sqrt{D[R]^2 + D[S]^2}} + \frac{D[R]}{\sqrt{D[R]^2 + D[S]^2}} U_R - \frac{D[S]}{\sqrt{D[R]^2 + D[S]^2}} U_S \quad \rightarrow$$

$$\bar{M} = \beta - \alpha_R U_R + \alpha_S U_S = \beta - \boldsymbol{\alpha}^T \mathbf{U} \tag{3.13}$$

From this normalized expression it is seen that the elementary reliability index can be interpreted as the distance from the origin (mean value point) to the limit state surface in the standardized u-space, Fig. 3.5. This result is valid for all safety margins linear in a set of basic normally distributed variables.

The point closest to the origin in u-space

$$\mathbf{u}^* = \beta \boldsymbol{\alpha} \tag{3.14}$$

is called the most likely failure point, also known as the β-point or the design point.

3.3.2 R and S non-normally distributed

If R and S are non-normally distributed the failure probability may be computed by the two-dimensional integral in Eq. 3.10. As before, a transformation into a standard u-space can be performed. However, now the transformation becomes non-linear

$$\{R, S\} = \{T_R(U_R), T_S(U_S)\} \tag{3.15}$$

and the limit state surface in u-space becomes curved.

FORM is required before the simulation takes place - effectively restricting the MCS-method to the class of analytical problems where FORM is applicable. Importance sampling can significantly reduce the variance of the estimator - theoretically to zero - if the sampling density is well located. This immense impact also implies that a poorly chosen sampling density can lead to very biased results, even for large but finite sample sizes.

In some recently suggested MCS approaches, adaptive sampling densities are used [Bucher, 1988]. During the sampling, the density is updated and changed according to some decision rule. The scheme has, thereby, a strong resemblance to the process of finding the most likely failure point in FORM. For some (ill-conditioned) problems, the result depends on the starting point, and the globally most likely failure point may not be identified. Adaptive sampling procedures and FORM/SORM have, therefore, some common pitfalls.

It appears that the use of FORM/SORM and advanced MCS methods does require some caution, insight, and experience. This is not different from many other engineering analysis methods such as, for example, technical beam theory and finite element methods, where in the latter case, the user must know which type of elements and which mesh size will lead to accurate results. If probability computation methods are used with foresight and understanding, the methods are powerful and can provide reliable results.

MCS and FORM/SORM methods for random variable reliability problems are described in detail in the following Secs. 3.3 and 3.4.

3.2.2 Random process reliability models

A random process reliability problem is defined by a safety margin $\min_{t\in[0,T]} G(\mathbf{X},\mathbf{Y}(t))$, where G is a failure function, $\mathbf{X}$ is a set of random variables, $\mathbf{Y}(t)$ is a set of random processes, and t is time. T is a reference period of time, typically the anticipated lifetime of the structure. The failure probability is generally a first outcrossing probability. In case of a scalar process, the problem reduces to an upcrossing problem.

In the simplest case, the failure surface in y-space is deterministic and time-independent. Even for this case almost no exact results for the first crossing probability is available. The probability can be estimated by simulation, see [Augusti et al, 1984; Hasofer et al, 1987]. A simple upper bound (or approximation) for the first crossing probability based on the mean outcrossing rate $\nu(t)$ can be used, see [Madsen et al, 1986; Melchers, 1987]. Whereas the failure probability in Eq. 3.1 was determined by a volume integral, the mean outcrossing rate is determined by a surface integral.

In the one-dimensional case of upcrossing, results for the upcrossing rate are available for a number of cases, see for example [Madsen et al, 1986; Melchers, 1987; Winterstein, 1988]. In the multi-dimensional case FORM/SORM type procedures based on exact or asymptotic results have been established for Gaussian processes, see [Veneziano et al, 1977; Ditlevsen, 1983; Hohenbichler and Rackwitz, 1986b; Wen, 1987]. Correspondingly, a directional simulation procedure for estimating the outcrossing rate is available in the Gaussian case [Ditlevsen et al, 1988]. For non-Gaussian processes, the outcrossing rate may be determined by the point crossing method or by the load coincidence model, see [Madsen et al, 1986; Wen and Chen, 1987; Bjerager et al, 1988a].

In the most recent developments the mean rate of vector processes outcrossing a safe set can be calculated using methods for time-independent reliability problems [Hagen and Tvedt, 1990;

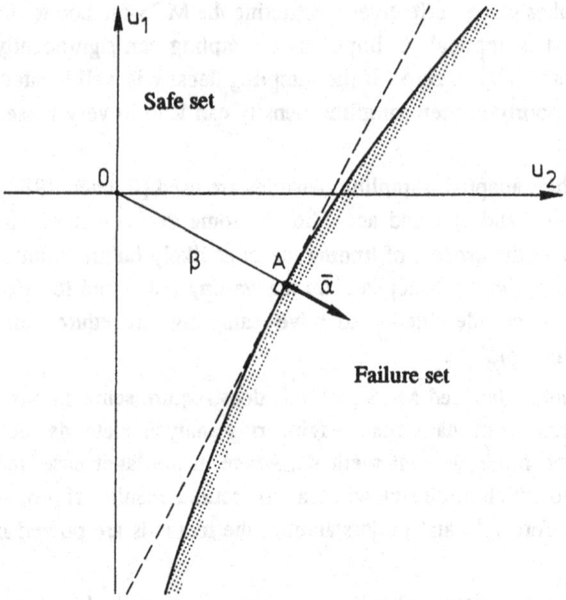

Fig. 3.6. Illustration of the geometric reliability index.

3.4 GAUSSIAN RELIABILITY PROBLEMS

In this section some important reliability problems defined in terms of an n-set of normally distributed basic variables $\mathbf{Y}^* = (Y_1, Y_2, \cdots, Y_n)$ of mean $E[\mathbf{Y}]$ and covariance matrix $\mathbf{C_Y}$ are considered. These Gaussian reliability problems are utilized in the reliability computation methods reviewed in the following sections.

The measure of reliability is the reliability index β related to the probability content p of the failure set, i.e.

$$\Phi(\beta) = P[\mathbf{Y} \in S] = P[G(\mathbf{Y}) > 0] = \int_{G(\mathbf{y})>0} f_{\mathbf{Y}}(\mathbf{y}) d\mathbf{y} = 1 - P[\mathbf{Y} \in F] \tag{3.17}$$

where S and F denote the safe set and failure set in y-space, respectively.

3.4.1 Linear safety margin

Because the class of n-dimensional normal distributions is closed with respect to in-homogeneous linear transformations, the linear safety margin

$$M = \mathbf{a}^T \mathbf{Y} + b \tag{3.18}$$

is normally distributed. M may then be referred to as a Gaussian safety margin. The corresponding reliability index is given as the elementary reliability index above, i.e.

$$\beta = \frac{E[M]}{D[M]} \tag{3.19}$$

*) In this section $\mathbf{Y}$ denotes *normally distributed (Gaussian)* basic variables. Generally, the basic random variables are denoted by $\mathbf{X}$.

where the expectation

$$E[M] = \mathbf{a}^T E[\mathbf{Y}] + b \tag{3.20}$$

and the variance

$$Var[M] = D[M]D[M] = \mathbf{a}^T \mathbf{C_Y a} \tag{3.21}$$

are computed from the second moment representation of **Y**.

It can be shown that linear safety margins are invariant to any in-homogeneous linear mapping which maps the y-space into another formulation space in the sense that the corresponding elementary reliability indices are invariant. An important transformation of this kind is the one transforming y-space onto a space of standardized and un-correlated variables **U**, whereby the safety margin can be brought into the form

$$M = \beta - \boldsymbol{\alpha}^T \mathbf{U}. \tag{3.22}$$

where $\boldsymbol{\alpha}$ is a unit outward normal vector to the limit state surface in u-space, Fig. 3.7.

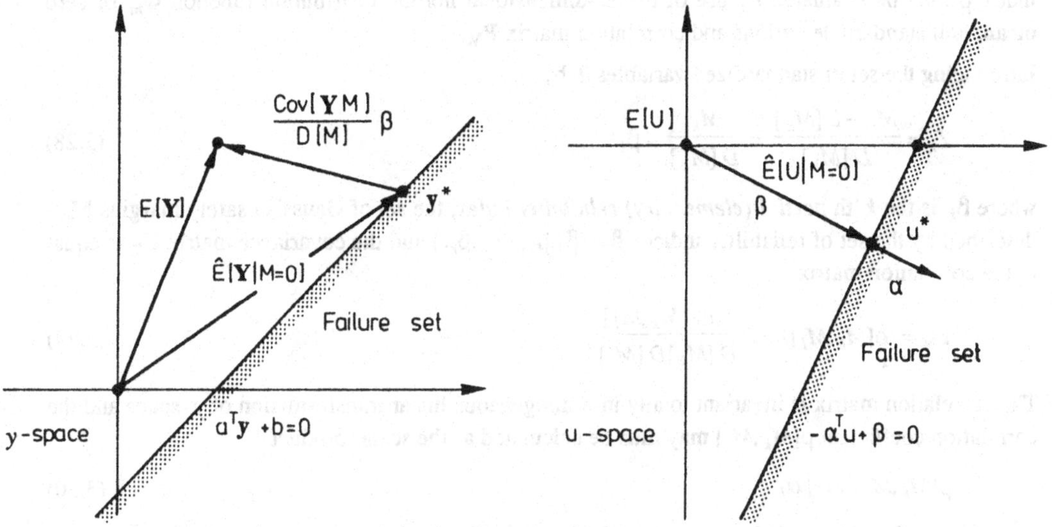

Fig. 3.7. Illustration of the most likely failure point on a hyperplane limit state surface in y-space and u-space.

The transformation between y-space and u-space is linear and is given as

$$\mathbf{U} = \mathbf{H}(\mathbf{Y} - E[\mathbf{Y}])\mathbf{H}^T \tag{3.23}$$

where the transformation matrix **H** is determined by

$$\mathbf{H}\mathbf{C_Y}\mathbf{H}^T = \mathbf{I} \tag{3.24}$$

where **I** is the identity matrix. The matrix **H** may be determined by a Cholesky decomposition. It can alternatively be determined by an eigen-value analysis of the covariance matrix. As regards the latter approach it is occasionally suggested that the dimension of the u-space can be reduced by neglecting contributions from the smaller eigen-values. However, the value of an eigen-value may not determine whether the contributions are significant for the reliability assessment (this depends on the actual G-function). A thorough sensitivity analysis should in general be employed.

3.4.2 Sets of linear safety margins

Let an m-set of linear safety margins be given as

$$\mathbf{M}=\mathbf{A}^T\mathbf{Y}+\mathbf{b} \tag{3.25}$$

where the matrix **A** and the vector **b** contain constants. **M** is then an m-dimensional normally distributed vector with mean

$$E[\mathbf{M}]=\mathbf{A}^T E[\mathbf{Y}]+\mathbf{b} \tag{3.26}$$

and covariance matrix

$$\mathbf{C_M}=\mathbf{A}^T\mathbf{C_Y}\mathbf{A} \tag{3.27}$$

When a reliability problem is defined by an m-set of Gaussian safety margins **M** the reliability index β may be evaluated by use of the m-dimensional normal distribution function Φ_m of zero mean, unit standard deviations and correlation matrix $\mathbf{P_M}$.

Introducing the set of standardized variables **Z** by

$$Z_k=\frac{M_k-E[M_k]}{D[M_k]}=\frac{M_k}{D[M_k]}-\beta_k \tag{3.28}$$

where β_k is the k'th *partial (elementary) reliability index*, the set of Gaussian safety margins **M** is described by the set of reliability indices $\boldsymbol{\beta}=(\beta_1,\beta_2,\cdots,\beta_m)$ and the covariance matrix $\mathbf{C_Z}$ is equal to the correlation matrix

$$\mathbf{P_M}=\{\rho[M_k,M_l]\}=\{\frac{Cov[M_k,M_l]}{D[M_k]D[M_l]}\} \tag{3.29}$$

The correlation matrix is invariant to any in-homogeneous linear transformation of y-space and the correlation coefficient $\rho[M_k,M_l]$ may thus be calculated as the scalar product

$$\rho[M_k,M_l]=\boldsymbol{\alpha}_k^T\boldsymbol{\alpha}_l \tag{3.30}$$

where $\boldsymbol{\alpha}_k$ is the unit normal vector to the k'th hyperplane in u-space corresponding to the vector defined in Eq. 3.22, Fig. 3.8. It is seen that the correlation coefficient $\rho[M_k,M_l]$ can be interpreted geometrically as the cosine of the angle between the two normal vectors in u-space.

Considering a *convex polyhedral safe set* in y-space, Fig. 3.8, the reliability index is given by

$$\Phi(\beta)=\Phi_m(\boldsymbol{\beta};\mathbf{P_M}) \tag{3.31}$$

on applying the *series system* formula

$$1-p=P[\min_{k=1}^{m}\{M_k\}>0]=P[\bigcap_{k=1}^{m}(M_k>0)] \tag{3.32}$$

Correspondingly, the *parallel system* formula

$$1-p=P[\max_{k=1}^{m}\{M_k\}>0]=P[\bigcup_{k=1}^{m}(M_k>0)] \tag{3.33}$$

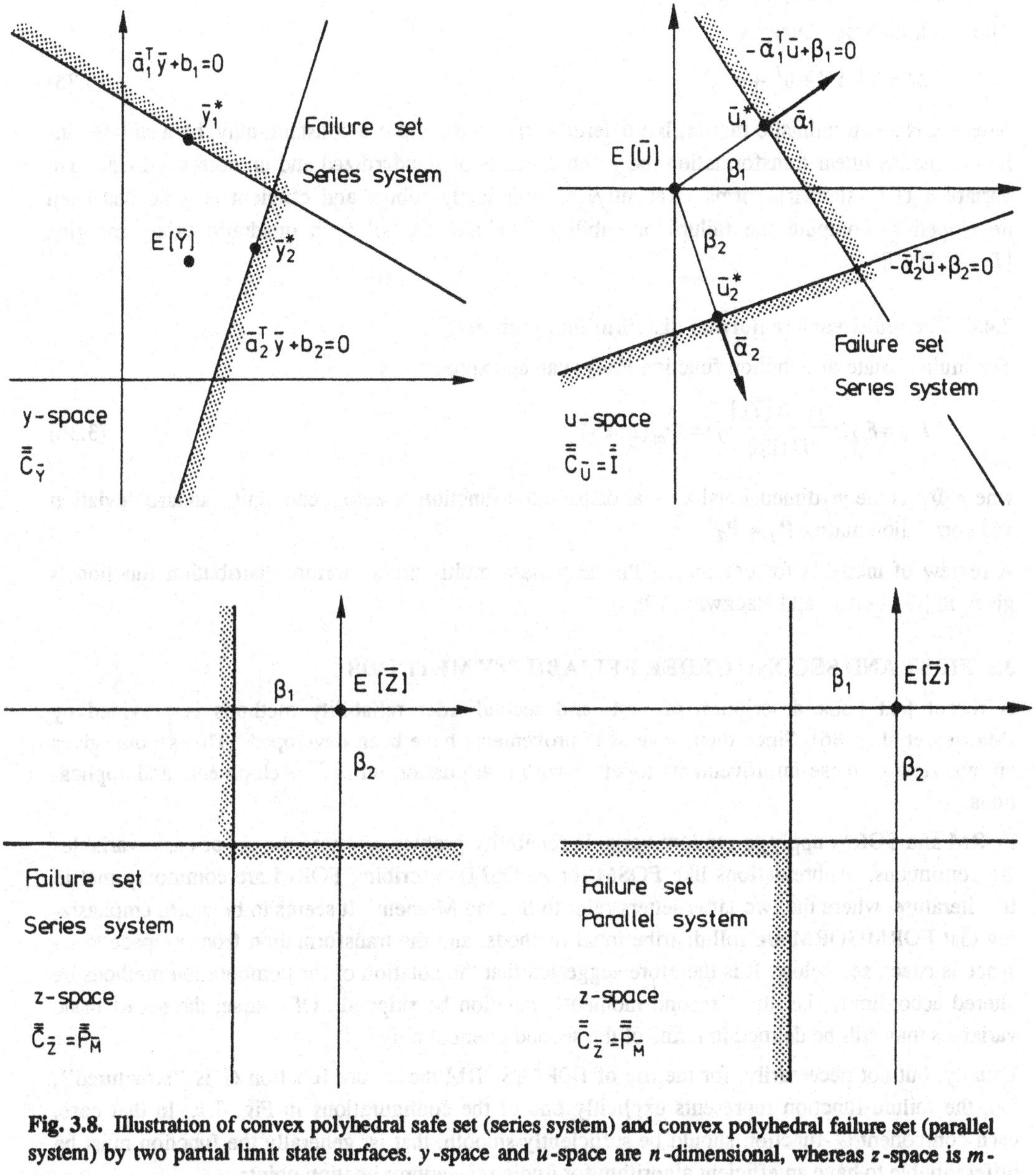

Fig. 3.8. Illustration of convex polyhedral safe set (series system) and convex polyhedral failure set (parallel system) by two partial limit state surfaces. *y*-space and *u*-space are *n*-dimensional, whereas *z*-space is *m*-dimensional (here two-dimensional). The convex polyhedral failure set is marked only in *z*-space.

implies

$$\Phi(\beta) = 1 - \Phi_m(-\beta; \mathbf{P_M}) \tag{3.34}$$

for a *convex polyhedral failure set* in *y*-space. To compute the reliability index, the multi-variate normal distribution function Φ_m has to be evaluated.

3.4.3 Quadratic safety margins

The quadratic safety margin

$$M = \mathbf{Y}^T \mathbf{A}\mathbf{Y} + \mathbf{b}^T + c \tag{3.35}$$

where **A** is a deterministic matrix, **b** a deterministic vector and c a constant, may, by a suitable in-homogeneous linear transformation, be given in terms of standardized and un-correlated Gaussian variables **U** in standard forms. Recently, a numerically robust and efficient scheme has been developed to compute the failure probability $P[M \leq 0]$ for all such quadratic safety margins [Tvedt, 1990].

3.4.4 The multi-variate normal distribution function

The multi-variate distribution function $F_{\mathbf{Y}}(\mathbf{y})$ can be expressed as

$$F_{\mathbf{Y}} = F_{\mathbf{Z}}(\{\frac{y_k - E[Y_k]}{D[Y_k]}\}) = \Phi_m(\mathbf{z}; \mathbf{P}_{\mathbf{Z}}) \tag{3.36}$$

where Φ_m is the m-dimensional normal distribution function of zero mean, unit standard deviation and correlation matrix $\mathbf{P}_{\mathbf{Y}} = \mathbf{P}_{\mathbf{Z}}$.

A review of methods for computing this important multi-variate normal distribution function is given in [Gollwitzer and Rackwitz, 1988].

3.5 FIRST AND SECOND ORDER RELIABILITY METHODS

A recent text book description of first- and second-order reliability methods is provided by [Madsen et al, 1986]. Since then, several improvements have been developed. This section gives an overview of these improvements together with a discussion on the developments and applications.

FORM and SORM apply to random variable reliability problems, where the set of basic variables are continuous. Abbreviations like FOSM (or AFOSM) describing FORM are commonly met in the literature, where the two latter letters refer to Second Moment. It seems to be worth emphasizing that FORM/SORM are full distributional methods, and the transformation from x-space to u-space is exact, see below. It is therefore suggested that the notation of the computation methods be altered accordingly, i.e. the "second moment" notation be skipped. Of course, the set of basic variables may still be defined in terms of the second moment only.

Usually, but not necessarily, for the use of FORM/SORM the failure function G is "structured", i.e., the failure function represents explicitly one of the configurations in Fig. 3.1. In that case, each component G-function should be sufficiently smooth, that is, generally the function must be differentiable to have an efficient algorithm for finding the approximation points.

A probability computation by FORM/SORM consists of three steps, 1) the *transformation* of **X** into the standard normal vector **U**, 2) an *approximation* of the failure surface in u-space, and 3), a *computation* of the failure probability corresponding to the approximating failure surfaces. The three steps are described in detail in the following sections.

3.5.1 Transformation of the basic random variables

The set of n basic random variables **X** is transformed into a set of q independent and standardized (zero mean, unit standard deviation), normally distributed variables **U**. Such a transformation is always possible for continuous random variables; in fact, the number of possible transformations is infinite. A particularly simple transformation can be used when the basic variables are mutually independent, with a cumulative distribution function F_{X_i}. Each variable can then be transformed separately with the transformation

$$\Phi(u_i) = F_{X_i}(x_i) \quad \rightarrow \quad u_i = \Phi^{-1}[F_{X_i}(x_i)] \tag{3.37}$$

where Φ is the standard normal distribution function. The transformation is illustrated in Fig. 3.9. Experience has shown that this transformation leads to relatively good approximations for the failure probability.

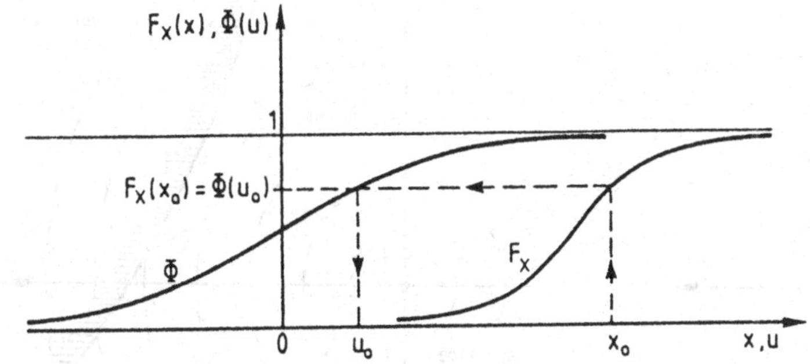

Fig. 3.9. The transformation of x-variables into u-variables for a single basic variable.

For dependent random variables an analogous transformation often referred to as the Rosenblatt transformation can be used, [Hohenbichler and Rackwitz, 1979]. In this case, a successive conditioning of the stochastically dependent variables leads to a transformation of the form

$$\begin{aligned} \Phi(u_1) &= F_{X_1}(x_1) \\ \Phi(u_2) &= F_{X_2 \mid X_1}(x_2 \mid x_1) \\ \Phi(u_3) &= F_{X_3 \mid X_2, X_1}(x_3 \mid x_2, x_1) \\ &\ldots \end{aligned} \tag{3.38}$$

A reliability model with stochastic dependencies will often be naturally defined in the successive way of Eq. 3.38. However, other transformations to the standard space can be used. For the cases of incomplete information with given marginal distributions and a correlation matrix for the set of physical variables **X**, the Nataf model can be used [Der Kiureghian and Liu, 1986]. When only marginal and joint moments for **X** are known, the Hermite moment transformation in [Winterstein and Bjerager, 1987; Winterstein, 1988] may be a convenient choice.

It is noted that, in general, the dimension q of u-space is not necessarily equal to the dimension of x-space, and q can be either smaller than, equal to, or larger than n. For example, if two or more of the basic variables are fully dependent, the dimension of u-space is smaller than the dimension of x-space. On the other hand, dependencies may be modeled by a number of independent variables leading to a transformation with $q \geq n$.

The one-to-one transformation of x-space on u-space maps the failure surface $\{\mathbf{x} \mid G(\mathbf{x})=0\}$ into a corresponding surface $\{\mathbf{u} \mid g(\mathbf{u})=0\}$ in u-space, see Fig. 3.10. The transformation preserves the probability mass p of the failure set $\{\mathbf{x} \mid G(\mathbf{x}) \leq 0\}$, i.e., in u-variables it is expressed as

$$p = P[g(\mathbf{U}) \leq 0] = \int_{g(\mathbf{u})\leq 0} f_{\mathbf{U}}(\mathbf{u}) d\mathbf{u} = \int_{g(\mathbf{u})\leq 0} \prod_{j=1}^{q} \frac{1}{\sqrt{2\pi}} \exp(-u_j^2/2) d\mathbf{u} \tag{3.39}$$

It is finally noted that the transformation to u-space is *exact* and that no approximation has been introduced by the transformation.

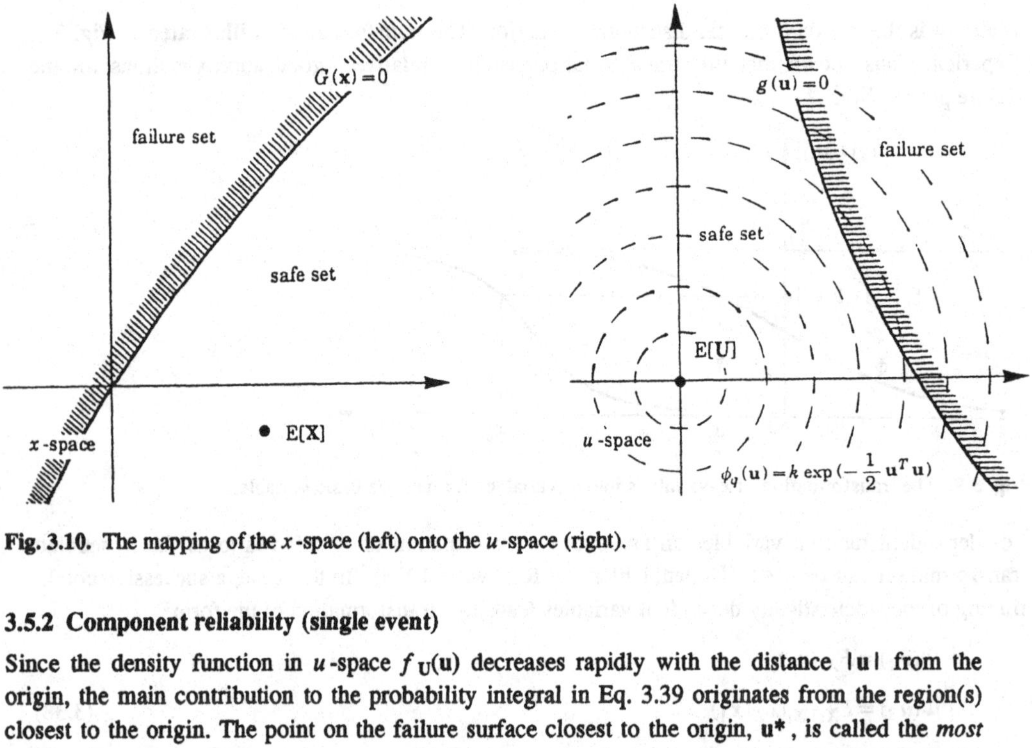

Fig. 3.10. The mapping of the x-space (left) onto the u-space (right).

3.5.2 Component reliability (single event)

Since the density function in u-space $f_{\mathbf{U}}(\mathbf{u})$ decreases rapidly with the distance $|\mathbf{u}|$ from the origin, the main contribution to the probability integral in Eq. 3.39 originates from the region(s) closest to the origin. The point on the failure surface closest to the origin, $\mathbf{u}^*$, is called the *most likely failure point*. It is also known as the *β-point* or the *design point*. In general there may be more than one such point.

The general idea in FORM/SORM is to approximate the failure surface in the most likely failure point by surfaces that are operational, i.e. for which it is possible to compute the corresponding failure probability (exactly or approximately). Therefore, using FORM/SORM essentially replaces a multi-dimensional integral by a mathematical programming problem for finding the most like failure point(s).

The basic case of FORM (and SORM) is concerned with a smooth, relatively flat component failure surface ($g(\mathbf{u})=0$) in u-space having one most likely failure point, Fig. 3.11 (left). Using FORM, the surface is approximated with a tangent hyperplane in this point, i.e. $g(\mathbf{u})$ is represented by its first order Taylor expansion at $\mathbf{u}^*$. The distance from the origin to the tangent hyperplane β

is called the *first order reliability index*, and it is equal to the minimum distance from the origin to the failure surface $|\mathbf{u}^*|$. The most likely failure point is conveniently expressed in terms of β and a unit directional vector $\boldsymbol{\alpha}$ as $\mathbf{u}^* = \beta\boldsymbol{\alpha}$, Fig. 3.11. The first-order approximation to the probability p is equal to the probability mass outside the approximating tangent hyperplane, i.e.

$$p \approx p^{FORM} = \Phi(-\beta) \tag{3.40}$$

A SORM approximation is obtained by approximating the failure surface in $\mathbf{u}^*$ by a second order surface, e.g. as defined by the second order Taylor expansion of $g(\mathbf{u})$, or by a curvature-fitted or point-fitted hyperparabolic surface, see [Fiessler et al, 1979; Breitung, 1984; Der Kiureghian et al, 1987]. The second order approximation to the probability p is then given as the probability content outside the second order surface, see Fig. 3.11 (left).

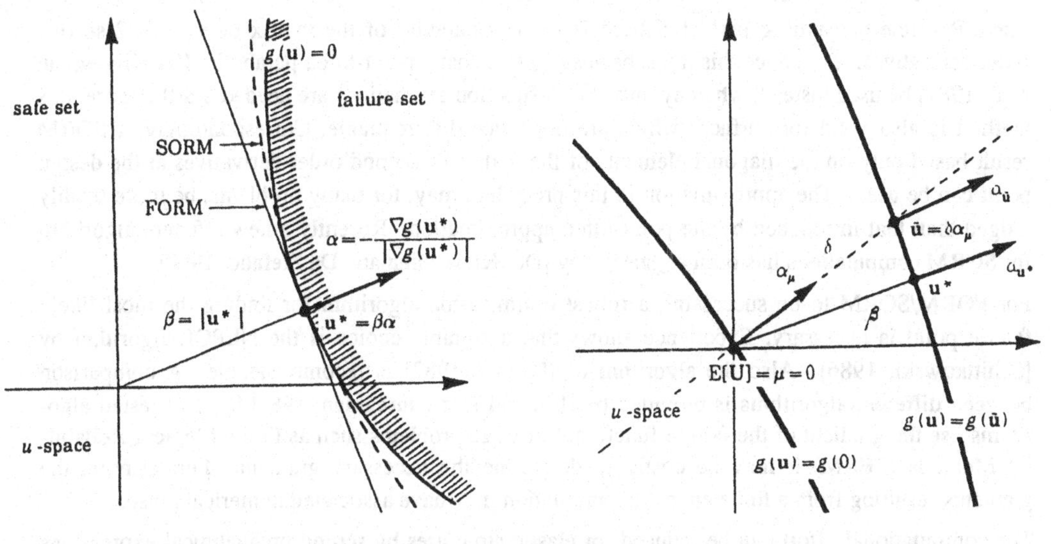

Fig. 3.11. Left: FORM/SORM approximation to smooth, relatively flat component surface. Right: Mean-based determination of an approximation $\bar{\mathbf{u}}$ to the most likely failure point $\mathbf{u}^*$.

SORM component reliability has recently been solved completely. The first thorough study on SORM was [Fiessler et al, 1979], where second-order Taylor expansions as well as curvature-fitted second order surfaces were considered. However, not before the important asymptotically exact result for parabolas was derived by [Breitung, 1984] was SORM considered a practical method. The second order result to p reads

$$p \approx p^{SORM}_{asymptotic} = \Phi(-\beta)\prod_{j=1}^{q-1}(1+\beta\kappa_j)^{-1/2} \tag{3.41}$$

where κ_j are the $q-1$ principal curvatures in $\mathbf{u}^*$. The sign convention is such that the curvatures are negative when the surface curves towards the origin. Improvements to this asymptotic result have been suggested. However, the final breakthrough for SORM did not come until recently, when an exact and numerically very feasible result was derived for a parabola [Tvedt, 1988], i.e.,

$$p \approx p^{SORM}_{parabolic} = \phi(\beta)\mathrm{Re}\left[i\left(\frac{2}{\pi}\right)^{1/2} \int_{t=0}^{i\infty} \frac{\exp\{(t+\beta)^2/2\}}{t} \left[\prod_{j=1}^{q-1}(1-t\kappa_j)^{-1/2}\right] \mathrm{d}t \right] \tag{3.42}$$

where ϕ is the standard normal density function and i is the imaginary unit. The one-dimensional integral can be efficiently evaluated by a saddle point integration method. This important result has significantly expanded the set of reliability problems that can accurately be handled by SORM. For example, experience has shown that exact parabolic SORM can also give very good accuracy for mid-range probabilities (0.1-0.9, say).

The exact formula for the parabola is presently being extended to cover all quadratic forms of Gaussian variables, whereby a general, second-order Taylor expansion SORM is now feasible [Tvedt, 1990]. However, it should be noted that where the curvature-fitted parabola is based solely on geometrical properties of the failure surface and as such is invariant with respect to the choice of failure function G, this is not the case for a second order Taylor expansion. Because of this invariance problem, the parabolic SORM may be preferable.

The CPU-time consuming part of SORM is the computation of the matrix of $n(n-1)/2$ second order derivatives. To reduce this it has been suggested that a point-fitted parabolic [Der Kiureghian et al, 1987] be used instead, whereby only $2n$ G-function evaluations are needed. Furthermore, this method is also valid for surfaces which are not twice differentiable. Correspondingly, a SORM result based only on the diagonal elements of the matrix of second order derivatives at the design point can be used. The approximation in this procedure may, for many problems, be more readily judged than that introduced by the point-fitted approximation. Recently a new efficient algorithm for SORM computations has been suggested by [Der Kiureghian and De Stefano, 1990].

For FORM/SORM to be successful, a robust optimization algorithm for finding the most likely failure point is necessary. Experience shows that a suitable choice is the NLPQL-algorithm by [Schittkowski, 1986]. Also the algorithm of [Powell, 1982] is recommendable. A comparison between different algorithms is provided by [Liu and Der Kiureghian, 1988b]. Most tested algorithms use the gradient of the failure function. For large problems such as Finite Element Reliability Methods (FERM) it may be costly to determine the necessary gradients. Furthermore, the gradients resulting from a finite element computation may have associated numerical noise.

The computational effort can be reduced for elastic structures by setting up analytical expressions for the gradients based, for example, on derivatives of the element stiffness matrices [Der Kiureghian, 1985; Liu et Der Kiureghian, 1988a]. In these computations results from structural sensitivity studies can be utilized, see [Arora, 1988]. Alternatively, the computation effort can be reduced by the use of omission sensitivity factors, see the discussion in Sec. 3.3.6.

One method of reducing the computational effort is to use - as the expansion point - an approximation $\mathbf{u}$ to the most likely failure point $\mathbf{u}^*$. One approach is known as the advanced mean-value FORM suggested by [Wu and Wirsching, 1987]. It has been used within probabilistic finite element analysis to find the probability distribution of a response quantity [Cruse et al, 1988; Wu et al, 1989]. It requires only one gradient computation, and it is a non-iterative scheme as regards the determination of the expansion point, i.e., there is no problem of convergence. The scheme is schematically shown in Fig. 3.11 (right) in u-space.

The choice of an expansion point in the tail of the distributions, as an extension to pure mean-based reliability, was originally suggested by [Paloheimo and Hannus, 1974]. In the advanced mean-value FORM it was suggested that the expansion point $\mathbf{u}$ be chosen by use of the normalized mean-value gradient vector (with opposite sign) $\boldsymbol{\alpha}_\mu$, Fig. 3.11 (right), such that - for chosen δ - the mean-value FORM approximation reads

$$P[g(\mathbf{U}) \leq g(\tilde{\mathbf{u}})] \approx P[g(\mathbf{U}) \leq g(\tilde{\mathbf{u}})]^{FORM} = \Phi(-\delta) \tag{3.43}$$

where $\tilde{\mathbf{u}} = \delta \boldsymbol{\alpha}_\mu$. If the probability distribution for $g(\mathbf{U})$ is sought, different values of δ are used. δ itself is an approximation to the corresponding first order reliability index β, see Fig. 3.11 (right). If the failure probability $P[g(\mathbf{U}) \leq 0]$ is sought, the zero-surface is found by iteration on δ. It is noted that the mean value in x-space typically does not map into the mean value in u-space.

For the case shown in Fig. 3.11 (right) the mean-value FORM gives a good result. However, it is easy to construct cases where this is not the case. Furthermore, when used in a reliability context, with iteration for the zero-surface, it is worth mentioning that the approach is not failure function invariant. Thus, care should be taken when mean-value FORM is applied.

The method can be improved (and the quality of the approximation may be judged) if the directional vector to the expansion point is based on more than one gradient. For example, the directional vector to the expansion point can be taken as

$$\boldsymbol{\alpha}(\delta) = \frac{\delta_0 - \delta}{\delta_0} \boldsymbol{\alpha}_\mu + \frac{\delta}{\delta_0} \boldsymbol{\alpha}_0 \tag{3.44}$$

where $\boldsymbol{\alpha}_0$ is a unit vector valid for the value $\delta = \delta_0$. Ideally, if a full FORM analysis for the value δ_0 is carried out, $\boldsymbol{\alpha}_0$ should be taken as the correct outward unit vector $\boldsymbol{\alpha}_{\mathbf{u}^*}$ to the design point. Alternatively, the normal vector $\boldsymbol{\alpha}_{\tilde{\mathbf{u}}}$ at $\mathbf{u} = \delta_0 \boldsymbol{\alpha}_\mu$ can be used. Of course, with an increased computational effort, more gradient vectors can be introduced in an attempt to improve the estimate for an appropriate expansion point.

Another way of reducing the number of G-function evaluations in a FORM-based approach is to use a response surface method (RSM). In particular, linear and quadratic surfaces are appropriate. The RSM methods are discussed in a subsequent section.

3.5.3 Parallel system reliability (small intersection)

If the failure surface in u-space is *singular* at the most likely failure point, an approximation to the surface is obtained by limiting tangent hyperplanes in this point. A singular most likely failure point is a typical result for a *parallel system*, see Fig. 3.12 (left). For the parallel system considered here it is assumed that $g(\mathbf{u}) > 0$, i.e., it is a *small* intersection. If $g(\mathbf{u}) < 0$, the system is converted to, and treated as, a series system by changing the sign on each of the single mode failure functions.

The most likely failure point on the parallel system failure surface is $\mathbf{u}^*$, Fig. 3.12 (left). The g-functions being equal to zero in this point are referred to as the *active* constraints. To further improve the approximation to the failure surface, a non-active failure function may be linearized at the point closest to the origin, which is both on this surface and on the system failure surface, see point $\mathbf{u}_h^*$ in Fig. 3.12 (left). The joint design points are found as the solution to a mathematical programming problem. Also, for this purpose, the NLPQL-algorithm may be used.

Assume k failure functions are linearized at the relevant approximation points, and the partial first order reliability indices β_i and the correlation coefficients $\rho_{ij} = \boldsymbol{\alpha}_i^T \boldsymbol{\alpha}_j$ for the set of approximating, linear safety margins are computed. The first order approximation to the failure probability of the parallel system is

$$p \approx p^{FORM} = \Phi_k(-\boldsymbol{\beta}, \mathbf{P}) \tag{3.45}$$

where $\boldsymbol{\beta}$ is the set of partial first order reliability indices, and Φ_k is the k-dimensional multivariate normal distribution function of zero means, unit variance and correlation matrix **P** determined by the correlation coefficients between the approximating safety margins. A robust and efficient evaluation method for the multi-normal distribution function is needed. The latest development on evaluation methods is documented in [Gollwitzer and Rackwitz, 1988].

An asymptotic second order approximation to the parallel system failure probability is of the form

$$p \approx p^{SORM} = \Phi_k(-\boldsymbol{\beta}, \mathbf{P})\,[\det(\mathbf{I}-\mathbf{D})]^{-1/2} \tag{3.46}$$

where **D** is a matrix containing second order derivatives of the active constraint in **u***, and **I** is the identity matrix of the same dimension, [Hohenbichler, 1984].

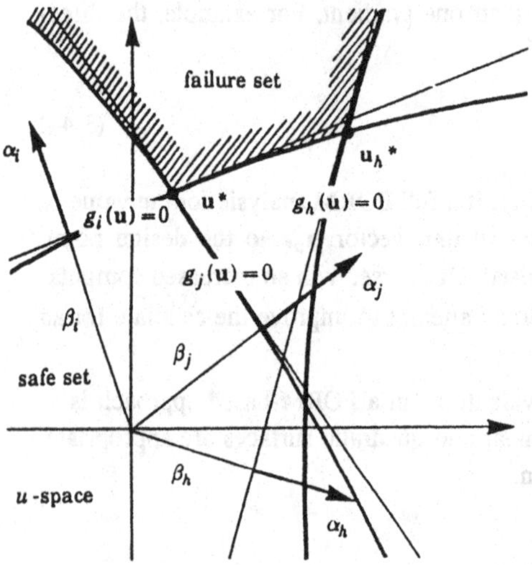

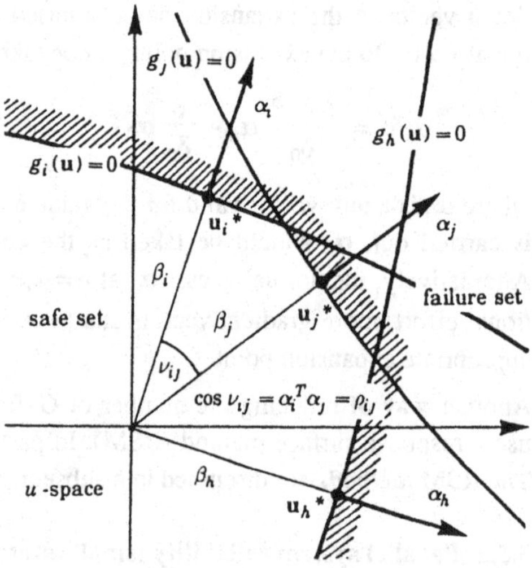

Fig. 3.12. Left: Small intersection (parallel system) defined by $m=3$ components. Right: Large intersection (series system) defined by $m=3$ components.

3.5.4 Series system reliability (large intersection)

It can be relevant to apply a multiple points approximation for a highly non-flat failure surface in u-space. A typical example of this is a *series system*, Fig. 3.12 (right), in which each partial failure surface corresponding to the components of the series system is approximated individually. Here it is assumed that $g(\mathbf{u})>0$ for the series system, i.e., that it is a *large* intersection. If $g(\mathbf{u})<0$, the system is converted to, and treated as, a parallel system by changing the sign on each of the single mode failure functions. If the most likely failure point on a partial failure surface falls inside the failure set for the series system, the corresponding failure function can - if such a point exists - be linearized at the point closest to the origin, which is both on the partial failure surface and on the series system failure surface.

For each of the linearizations, the partial reliability index β_i and the corresponding outward unit vectors $\boldsymbol{\alpha}_i$ are determined. The correlation matrix **P** for the set of approximating linear safety margins are given by the correlation coefficients $\rho_{ij}=\boldsymbol{\alpha}_i^T\boldsymbol{\alpha}_j$, and the first order probability estimate for the series system is

$$p \approx p^{FORM} = 1 - \Phi_k(\boldsymbol{\beta}, \mathbf{P}) \tag{3.47}$$

Alternatively, for series systems the probability corresponding to the multiple points approximation can be bounded by use of the well-known second-order probability bounds [Ditlevsen, 1981; Madsen et al, 1986]. The bounds are close for small probabilities. A second-order probability estimate for a series system can be obtained by a using second order analysis on each of the individual failure surfaces as well as the joint sets (parallel systems) entering the probability bounds.

3.5.5 General systems reliability

A series system of parallel sub-systems can be directly treated as a series system using the second-order bounds. A parallel system analysis is carried out for each of the single mode events as well as for the pairwise joint events defined by two parallel sub-systems in parallel.

With FORM/SORM there is at present no direct way of treating parallel systems of series sub-systems, and instead such systems are reformulated into series systems of parallel sub-systems. This reformulation is only practicable for smaller systems. Correspondingly, general reliability network, as well as event and fault tree representations must be converted into a representation by a series system of parallel sub-systems, i.e. into a cut set representation.

3.5.6 Sensitivity measures

The probability in Eq. 3.1 is typically a function of a number of fixed variables, here referred to as parameters. These parameters may describe the probability distribution of **X** in terms of mean values, standard deviations and correlation coefficients, or the parameters may enter the failure function directly as non-random (fixed) basic variables. A parametric sensitivity analysis is an important part of a modern structural reliability analysis.

A parametric sensitivity factor γ_θ is defined as the derivative

$$\gamma_\theta = \frac{dp}{d\theta} \tag{3.48}$$

where θ is the parameter. Parametric sensitivity factors for alternative reliability measures may also be introduced, and with the relation $p = \Phi(-\beta_R)$ the relation

$$\frac{dp}{d\theta} = -\frac{1}{\phi(\beta_R)} \frac{d\beta_R}{d\theta} \tag{3.49}$$

holds between the factors for the generalized reliability index β_R and the failure probability p.

Parametric sensitivity factors have several useful applications. For example, the probability density function of the random variable $Y = G(\mathbf{X})$ can be formulated as a parametric sensitivity factor by

$$f_Y(y) = \frac{d}{dy} P[G(\mathbf{X}) - y \le 0] \tag{3.50}$$

In reliability-based optimization the derivatives are used by the optimization algorithm, and in reliability based-design the factors can be used to guide the design process. It is noted that the

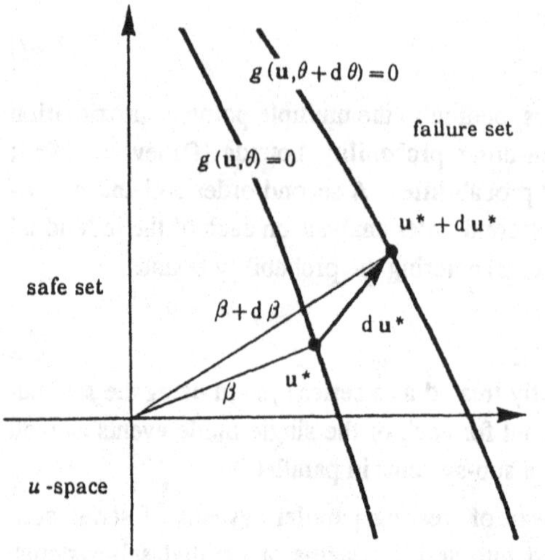

Fig. 3.13. Illustration of first order parametric sensitivity.

reliability index typically is "more linear" in the parameter θ than the failure probability, such that it is preferable to predict the changes in reliability due to a change in the parameter value by using the parametric sensitivity factor for the the reliability index rather than that for the failure probability.

By FORM, parametric sensitivity factors can be determined by relatively little extra computational effort, once the probability has been computed. As for the probability calculation in FORM, the computation of a parametric sensitivity $dp/d\theta$ is based on the most likely failure point $\mathbf{u}^*$ as well as on the derivative $d\mathbf{u}^*/d\theta$, Fig. 3.13. It can be shown that the derivative of the first order reliability index is

$$\frac{d\beta}{d\theta} = \boldsymbol{\alpha}^T \frac{d\mathbf{u}^*}{d\theta} = \frac{1}{|\nabla g(\mathbf{u}^*)|} \frac{\partial g(\mathbf{u}^*)}{\partial \theta} \qquad (3.51)$$

For a distribution parameter only the derivative of the transformation is needed, see [Madsen et al, 1986]. In [Hohenbichler and Rackwitz, 1986a] a number of asymptotic results are given. The asymptotic results assume the change of the unit directional vector $\boldsymbol{\alpha}$ to the most likely failure point in u-space is zero. For optimization purposes it is desirable to have derivatives in exact correspondence with the reliability. In [Bjerager and Krenk, 1987] the derivative of the most likely failure point and thereby the unit normal vector has therefore been derived. The derivative depends on the second order derivatives of the failure function in $\mathbf{u}^*$. In [Madsen, 1990] the results have been extended to cover parallel systems where the derivative of the joint design point is determined by a set of equations. Similarly, FORM sensitivity results can be constructed for series systems by deriving the multivariate normal distribution function in Eq. 3.47.

It is noted that only efficient results for first order parametric sensitivity factors exist. If the curvatures of the failure surface depend on the parameter θ, these factors may not be sufficiently accurate. For such cases, it may be necessary to develop more accurate methods for sensitivity analysis. An exact SORM parametric sensitivity may not be feasible since it requires third order derivatives, but some approximations can be established.

For independent random variables the components of the directional vector α to the most likely failure point can be taken as first order measures of the relative importance of the uncertainty in the basic variables. If a basic variable with a small α^2 value is replaced by its median, the first order reliability result will remain almost unchanged. The α^2 values therefore give an indication of the (first order) *uncertainty fraction* that can be associated with the corresponding random variable, and the quantity is commonly - but maybe not very precisely - referred to as an importance factor.

The unit directional vector α may be used to reduce the number of variables in the search for the most likely failure point [Der Kiureghian, 1985]. The components of the vector can be used to compute an approximation to the first order reliability index when one variable is substituted by, for example, its median value [Madsen, 1988; Igusa et al, 1988]. Furthermore, they can be used to compute *omission sensitivity* factors to determine somewhat optimal deterministic substitute values of non-important variables during the search for the most likely failure point, [Madsen, 1988; Bjerager and Arnbjerg-Nielsen, 1988]. This may be an important way of reducing the computational effort in a consistent FORM analysis.

For example, if the substitute value u_j^f of the jth variable in u-space is taken to be proportional to α_j, a set of optimal substitute values (in case the failure surface is linear) can be shown to be

$$u_j^f = \frac{1-(1-\sum\alpha_i^2)^{1/2}}{\sum\alpha_i^2}\beta\alpha_j \;\rightarrow\; \frac{1}{2}\beta\alpha_j \quad \text{for} \quad \sum\alpha_i^2 \rightarrow 0 \tag{3.52}$$

where the sum is taken over all substituted u-variables. Based on α values in the first iteration, less important variables are substituted by the fixed value. The value may be updated with respect to the current value of β as the iteration proceeds. Finally, a check computation of the full gradient vector should be performed. For dependent variables, each deterministic substitute value for a physical x-variable must be based on a set of α-values.

It is finally mentioned, that general omission sensitivity factors can be computed by solving conditional reliability problems of the form in Eq. 3.67 below, with $H(\mathbf{X})=X_i-x_i^f=0$, where x_i^f is the deterministic substitute value for X_i, typically taken as the median value of X_i. The case where several random variables are substituted with fixed values can be handled correspondingly.

3.5.7 Characteristics of FORM/SORM

FORM and SORM are analytical and approximate methods, and their accuracy is generally good for small probabilities, which are the situations for which the methods have been developed. Analytical properties enable the methods to yield relatively inexpensive sensitivity factors, and to be efficient for the computation of conditional reliability.

Generality: FORM/SORM are analytical probability computation methods, and the methods apply therefore to probability problems with certain (but few) analytical properties. The basic random variables must be continuous, and each (component) failure function must be continuous. With the optimization procedures presently used in most cases, the failure functions should also be smooth. Finally, the failure function must generally be given in a structured way.

Accuracy: The methods are approximate, but yield generally accurate results for practical purposes, in particular for small range probabilities ($10^{-3}-10^{-8}$, say) and a uni-modal joint density of the basic variables. Furthermore, with the recently established exact SORM, a good accuracy can be obtained for all probability levels for component failure functions.

Efficiency: For small order probabilities FORM/SORM are extremely efficient as compared to simulation methods, and as such are without competition, as regards CPU-time. The CPU-time for FORM is approximately linear in the number of basic variables n, and the additional CPU-time for a SORM computation grows approximately with n^2. Furthermore, the computation time is roughly linear in the number of components (or parallel sub-systems), for series systems. The absolute computation time depends on the time necessary to evaluate the failure functions. This time may in fact depend on the actual values of the basic variables. Extreme values may take longer, due to increased non-linearities in the problem. The CPU-time is independent of the probability level, assuming a constant time for evaluation of the failure functions.

Restrictions and further developments: When the failure surface is not sufficiently smooth, the approximation point, i.e. the most likely failure point, cannot be identified by efficient mathematical programming methods applying the gradients of the function. In this case it would be advantageous to fit the exact failure function by a differentiable function. Correspondingly, if the function is extremely CPU-time costly to evaluate, a simpler function may be used. Classical statistics suggest to use response surface methods (RSM). A brief review of the the use of RSM within structural reliability analysis is given in Sec. 3.7.

A fundamental requirement for a reliability problem to be analyzed by FORM/SORM is that the transformation from x-space to u-space exists, i.e., that the basic random variables are continuous. If this is not the case, assume that **X** can be divided into a set of continuous random variables and a set of discrete random variables. The reliability problem can then be solved by conditioning on the discrete variables, where each conditional reliability problem may be solved by FORM/SORM. However, the number of different outcomes for the discrete random variables can easily become very high, and this direct approach may not be practicable. Alternatively the continuous/discrete reliability problem may be solved by a conditional expectation simulation method, where the discrete variables are simulated, and FORM/SORM are used to solve the conditional reliability problems. Further research on the continuous/discrete reliability models and use of FORM/SORM is needed.

3.6 MONTE CARLO SIMULATION METHODS

MCS can be used in general for probability integration, or it can be used specifically to update FORM/SORM results to exact results. In this section an overview of recent developments within two classes of MCS for random variable reliability problems are given, namely 1) the zero-one indicator-based methods based on x-space, and 2) the semi-analytical, conditional expectation methods based on u-space. The latter methods are primarily used to check and update FORM/SORM results, whereas the former methods first of all are aimed at the class of non-analytical problems for which other methods can not be applied.

For use with the simulation methods there are less strict requirements on the analytical properties of the failure function, and functions of the non-structured, "black box" type can be used. Of course, structured failure functions can also be applied with simulation.

3.6.1 Zero-one indicator-based MCS

The zero-one indicator-based Monte Carlo Simulation is the basic MCS described in numerous papers and text books. The methods are, in particular, efficient for mid-range probabilities and become generally costly for the small probabilities typically encountered in structural and mechanical reliability. Recent presentations of the methods can be found in [Augusti et al, 1984; Ang and Tang, 1984; Melchers, 1987].

Let the failure probability be written as

$$p = \int_{\mathbf{x} \in \mathbf{R}^n} I[G(\mathbf{X}) \leq 0] \frac{f_{\mathbf{X}}(\mathbf{x})}{h_{\mathbf{X}}(\mathbf{x})} h_{\mathbf{X}}(\mathbf{x}) d\mathbf{x} \tag{3.53}$$

where I is an indicator function equal to one when the argument is true, and otherwise zero, and $h_{\mathbf{X}}(\mathbf{x})$ is the non-negative sampling density. By performing N simulations of the vector $\mathbf{X}$ with respect to $h_{\mathbf{X}}(\mathbf{x})$, p is estimated as the average of the sample values

$$p_i = I[G(\mathbf{x}_i) \leq 0] \frac{f_{\mathbf{X}}(\mathbf{x}_i)}{h_{\mathbf{X}}(\mathbf{x}_i)} \tag{3.54}$$

i.e.,

$$\hat{E}[P] = \frac{1}{N} \sum_{i=1}^{N} p_i \tag{3.55}$$

An estimate of the standard deviation $\hat{D}[P]$ on the estimator is given by

$$\hat{D}[P]^2 = \frac{1}{N(N-1)} \sum_{i=1}^{N} (p_i - \hat{E}[P])^2 \tag{3.56}$$

Assuming the number N to be sufficiently large the estimator for p can according to the central limit theorem be assumed to be normally distributed. This implies that an upper and lower bound on p can be established by

$$P\{\hat{E}[P] + \Phi^{-1}(\kappa)\hat{D}[P] \leq p \leq \hat{E}[P] - \Phi^{-1}(\kappa)\hat{D}[P]\} \approx 1 - 2\kappa \tag{3.57}$$

for $\kappa < ½$.

In structural reliability analysis p is typically of the order 10^{-4} or less. This implies that the sample size N in the described Monte Carlo simulation approach must be very large in order to obtain a sufficiently reliable estimate for p. In the mean it thus takes about $1/p_f$ simulations to obtain one outcome of $\mathbf{X}$ in the failure set.

In effect, only knowledge about whether or not the sampled value is in the failure set is used. In basic Monte Carlo simulation the sampling density is taken as the original density of $\mathbf{X}$, i.e., $h_{\mathbf{X}}(\mathbf{x}) = f_{\mathbf{X}}(\mathbf{x})$, and the sample value is then simply equal to the value of the indicator function.

3.6.2 Variance reduction techniques

Variance reduction techniques (VRT) may be applied to increase the efficiency of MCS. A number of procedures are suggested in the literature given above together with standard text books on simulation, for example [Rubenstein, 1981].

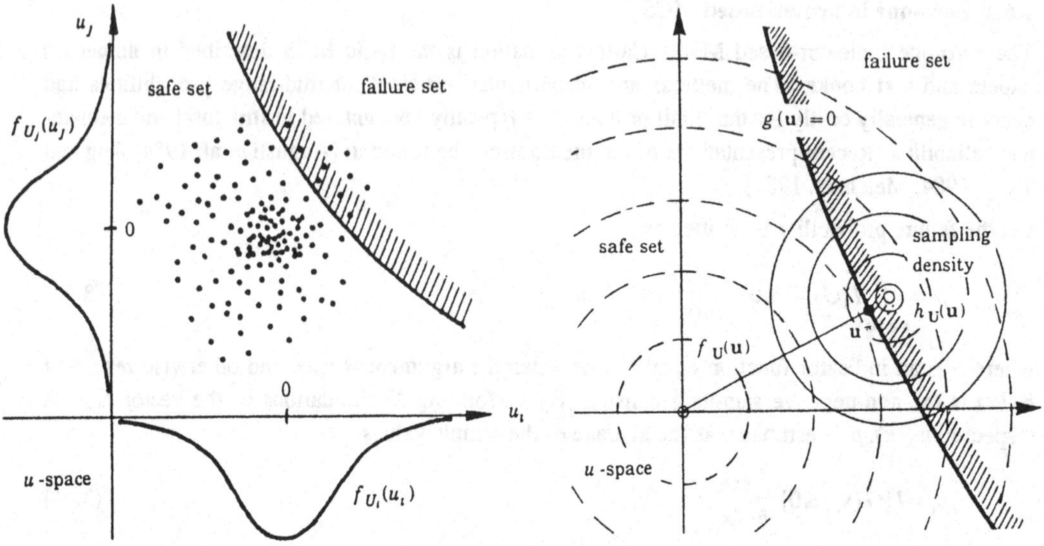

Fig. 3.14. Indicator-based Monte Carlo sampling in u-space (left), and importance sampling around the most likely failure point (right).

In structural reliability, in particular, *importance sampling* seems to be an attractive VRT, [Shinozuka, 1983; Harbitz, 1983]. The idea is to choose the sampling density $h_{\mathbf{X}}(\mathbf{x})$ to reduce the variance of the estimator. Theoretically, using Eq. 3.24 with $p_i = p$, the variance on the estimator can be reduced to zero by use of the sampling density

$$h_{\mathbf{X}}(\mathbf{x}) = I[G(\mathbf{x}) \le 0] \frac{f_{\mathbf{X}}(\mathbf{x})}{p} \tag{3.58}$$

This sampling density corresponds to exclusive sampling in the failure set according to the original density $f_{\mathbf{X}}(\mathbf{x})$. This is not practicable, of course, for the single reason that it requires a knowledge of the probability p, which is the aim of the simulation. However, the consideration can be used to establish suitable choices of the sampling density, such as those with sampling densities located in the neighbourhood of the most likely failure point, see [Melchers, 1987]. Such an importance sampling is illustrated in Fig. 3.14 (right), showing the standard normal density $f_{\mathbf{U}}(\mathbf{u})$ and the sampling density $h_{\mathbf{U}}(\mathbf{u})$ in u-space. As mentioned previously, methods for updating the sampling density during the simulation has recently been suggested, see [Karamchandani, 1987; Bucher, 1986]. More experience on the robustness of these approaches is needed.

Other variance reduction techniques may be applied. For example, stratified sampling and, in particular, Latin hypercube sampling has been suggested as an efficient method, [McKay et al, 1979; Iman and Helton, 1985]. In these sampling techniques the sampling space is divided into subsets of equal probability. In stratified sampling an outcome is generated in each sub-set, Fig. 3.15 (left). In Latin hypercube sampling the number of samples are reduced to represent each subset of each (independent) random variable only once, Fig. 3.15 (right). Latin hypercube sampling is relatively efficient for generating the mean and standard deviation of $G(\mathbf{X})$, whereas no efficiency gain is generally obtained for small order failure probabilities in structural reliability.

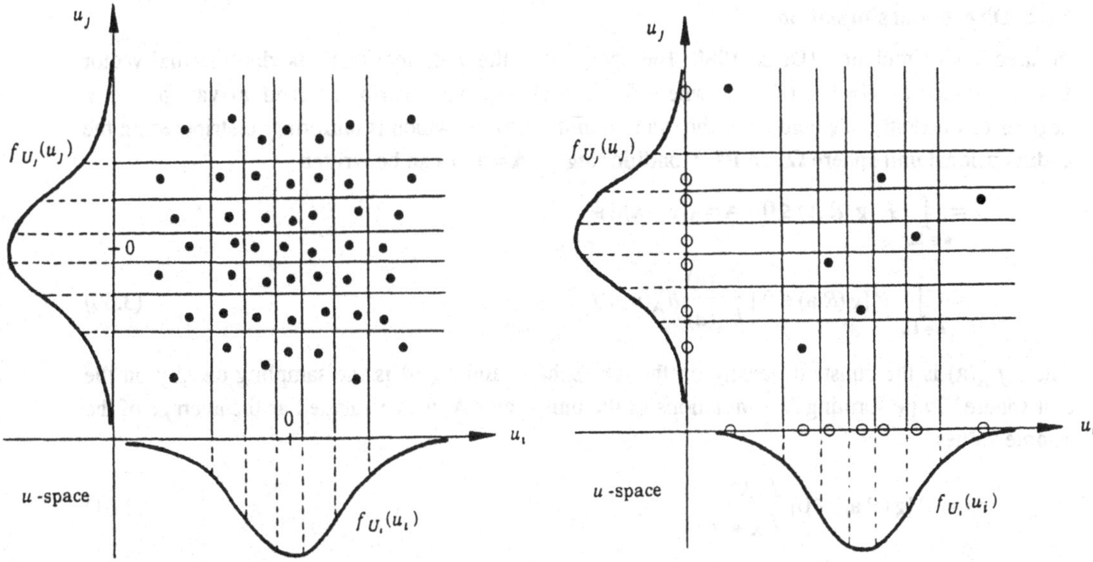

Fig. 3.15. Stratified sampling in u-space (left), and Latin hypercube sampling (right).

3.6.3 Characteristics of indicator-based MCS

Generality: The direct Monte Carlo simulation is completely general, and applies to any distribution of the basic random variables, including discrete random variables. Furthermore, there is no restrictions on the failure functions - only the sign of the failure function is used.

Accuracy: For sample size $N \to \infty$, the estimated probability converges to the exact result. For finite sample sizes, uncertainty estimates on the results are provided.

Efficiency: As a general rule, the CPU-time grows approximately linearly with $1/p$ and linearly with q for a given coefficient of variation of the estimator. The absolute computation time depends on the time necessary to evaluate the failure functions. For small range probabilities the method is generally very expensive in CPU-time. As a rule of thumb, the necessary sample size to get a probability estimate with good confidence is around $100/p$.

3.6.4 Conditional expectation MCS

One general variance reduction technique is that of conditional expectation. Let $\mathbf{X}=\{\mathbf{Y},\mathbf{Z}\}$ be the set of random variables. The failure probability p can then be expressed as

$$p = \int_{\mathbf{y} \in \mathbf{R}_r} P[G(\mathbf{Z},\mathbf{Y}) \leq 0 \mid \mathbf{Y}=\mathbf{y}] \frac{f_{\mathbf{Y}}(\mathbf{y})}{h_{\mathbf{Y}}(\mathbf{y})} h_{\mathbf{Y}}(\mathbf{y}) d\mathbf{y} \tag{3.59}$$

If the conditional probability $P[G(\mathbf{Z},\mathbf{y}) \leq 0]$ is easy to compute, the failure probability may be efficiently estimated by generating outcomes of $\mathbf{Y}$ according to the sampling density $h_{\mathbf{Y}}(\mathbf{y})$, and then computing the conditional probability in Eq. 3.59 for each outcome. The variance of the estimator may be reduced, since part of the probability integration is done analytically.

Recently, two conditional expectation methods for structural and mechanical reliability computations have been formulated, namely 1) *directional simulation*, and 2) what will here be denoted *axis-orthogonal simulation*. The methods work in the transformed space, u-space.

3.6.5 Directional simulation

In directional simulation [Deak, 1980; Bjerager, 1988] the q-dimensional standard normal vector $\mathbf{U}$ is expressed as $\mathbf{U}=R\,\mathbf{A}$ ($R \geq 0$), where R^2 is a chi-squared distributed random variable of q degrees of freedom, independent of the random unit vector $\mathbf{A}$, which is uniformly distributed on the q-dimensional unit sphere Ω_q in $\mathbf{R}^q$. Conditioning on $\mathbf{A}=\mathbf{a}$, p can be written

$$p = \int_{\mathbf{a} \in \Omega_q} P[g(R\,\mathbf{A}) \leq 0 \mid \mathbf{A}=\mathbf{a}] f_{\mathbf{A}}(\mathbf{a}) d\mathbf{a}$$

$$= \int_{\mathbf{a} \in \Omega_q} P[g(R\,\mathbf{a}) \leq 0] \frac{f_{\mathbf{A}}(\mathbf{a})}{h_{\mathbf{A}}(\mathbf{a})} h_{\mathbf{A}}(\mathbf{a}) d\mathbf{a} \tag{3.60}$$

where $f_{\mathbf{A}}(\mathbf{a})$ is the constant density on the unit sphere, and $h_{\mathbf{A}}(\mathbf{a})$ is the sampling density on the unit sphere. By performing N simulations of the unit vector $\mathbf{A}$, p is estimated as the average of the sample values

$$p_i = P[g(R\,\mathbf{a}_i) \leq 0] \frac{f_{\mathbf{A}}(\mathbf{a}_i)}{h_{\mathbf{A}}(\mathbf{a}_i)} \tag{3.61}$$

An estimate for the uncertainty on the estimator can also be computed.

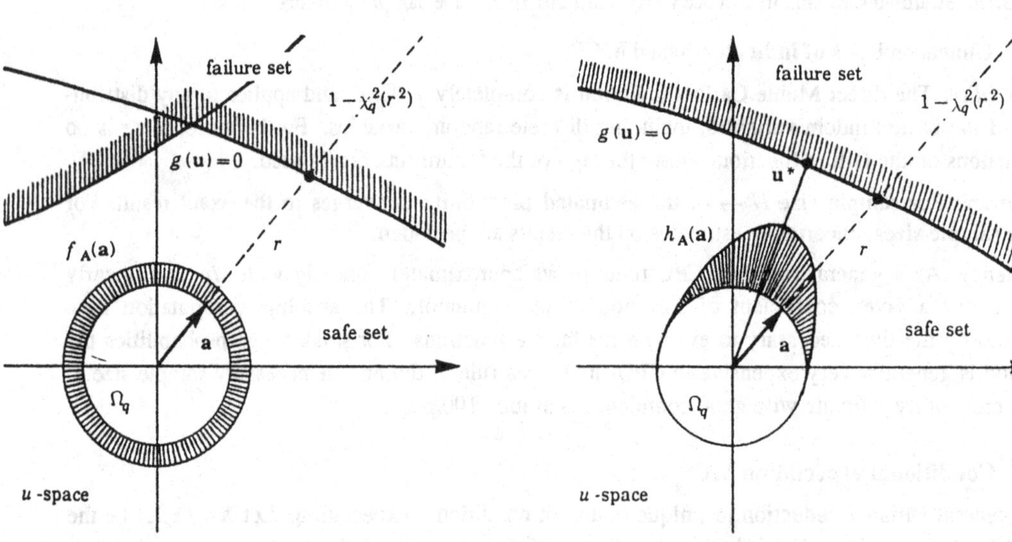

Fig. 3.16. Uniform directional simulation (left), and directional importance sampling (right).

The sample value is computed by finding the roots for r in $g(r\,\mathbf{a})=0$, $r \geq 0$, see Fig. 3.16. The smallest root is simply the distance from the origin to the failure surface in the direction of $\mathbf{a}$. If only one root r exists, the sample value is simply $1-\chi_q^2(r^2)$, where χ_q^2 is the chi-square distribution function. If several roots exist, the sample value becomes a sum of similar contributions.

Directional simulation with uniform directional sampling is built upon the symmetry of the u-space, and it is particularly efficient for failure surfaces close to spheres with a centre at the origin. It can be used as a stand-alone simulation method or in combination with FORM/SORM to update the approximate results to exact results. In the latter case, based on theoretical considerations such as the ones given with Eq. 3.58, suitable directional importance sampling densities for typical FORM/SORM surface approximations can be set up [Bjerager, 1988], Fig. 3.16 (right). For series systems, a stratified sampling density is used.

A special second-order reliability method for parallel systems based on directional sampling has been suggested [Lin and Der Kiureghian, 1987]. The method allows for an efficient determination of the distance from the origin to the (approximate) failure surface. Methods for adaptive sampling should be further investigated. Suggested procedures can be found in [Bucher, 1988, Karamchandani et al, 1989]. Presently, a directional simulation approach in x-space is being investigated [Ditlevsen et al, 1989].

3.6.6 Axis-orthogonal simulation

Axis-orthogonal simulation is aimed solely at updating FORM/SORM results [Fujita and Rackwitz, 1987; Hohenbichler and Rackwitz, 1988; Schall et al, 1988]. The q-dimensional coordinate system in u-space is rotated into a $\{\mathbf{V}, V_q\}$-system, where $\mathbf{V}$ contains a $(q-1)$-dimensional sub-space, and the v_q-axis is the axis through the origin and the most likely failure point for basic FORM/SORM approximations, see Fig. 3.17 (left). For a FORM/SORM approximation to a parallel system, the v_q-axis is parallel to the mean direction of the normal vectors of the limiting tangent hyperplanes in the singular most likely failure point, [Hohenbichler and Rackwitz, 1988], see Fig. 3.17 (right). Conditioning on $\mathbf{V}=\mathbf{v}$, p can be written

$$p = \int_{\mathbf{v} \in R^{q-1}} P[g(\{\mathbf{V}, V_q\}) \leq 0 \mid \mathbf{V}=\mathbf{v}] f_{\mathbf{V}}(\mathbf{v}) d\mathbf{v}$$

$$= \int_{\mathbf{v} \in R^{q-1}} P[g(\{\mathbf{v}, V_q\}) \leq 0] \frac{f_{\mathbf{V}}(\mathbf{v})}{h_{\mathbf{V}}(\mathbf{v})} h_{\mathbf{V}}(\mathbf{v}) d\mathbf{v} \tag{3.62}$$

where $f_{\mathbf{V}}(\mathbf{v})$ is the $(q-1)$-dimensional, standard normal density on the hyperplane $v_q=0$, and $h_{\mathbf{V}}(\mathbf{v})$ is a sampling density on the space. By performing N simulations of the $(q-1)$-dimensional vector $\mathbf{V}$ according to $h_{\mathbf{V}}(\mathbf{v})$, p is estimated as the average of the sample values

$$p_i = P[g\{\mathbf{v}_i, V_q\}) \leq 0] \frac{f_{\mathbf{V}}(\mathbf{v}_i)}{h_{\mathbf{V}}(\mathbf{v}_i)} \tag{3.63}$$

The sample value is computed by finding the roots for v_q in $g\{\mathbf{v}, v_q\}=0$, see Fig. 3.17. The smallest positive root is simply the distance in the positive v_q-direction from the plane $v_q=0$ to the failure surface parallel to the v_q-axis. If only one root $v_q \geq 0$ exists, the sample value is simply $1-\Phi(v_q)$. If several roots exist, the sample value becomes a sum of similar contributions.

The axis orthogonal sampling is not based on any particular property of the u-space. The sampling density $h_{\mathbf{V}}(\mathbf{v})$ is determined based on FORM/SORM results. As seen, the mean value of the sampling density may be shifted away from the origin (parallel system), and the standard deviations altered to somehow reflect the curvature in the approximation point, [Schall et al, 1988]. In the case of a singular approximation point (parallel system), concentrated curvatures are used. Alternatively, the sampling densities derived in [Bjerager, 1988] can be used. Improvement of the sampling schemes for systems may be necessary.

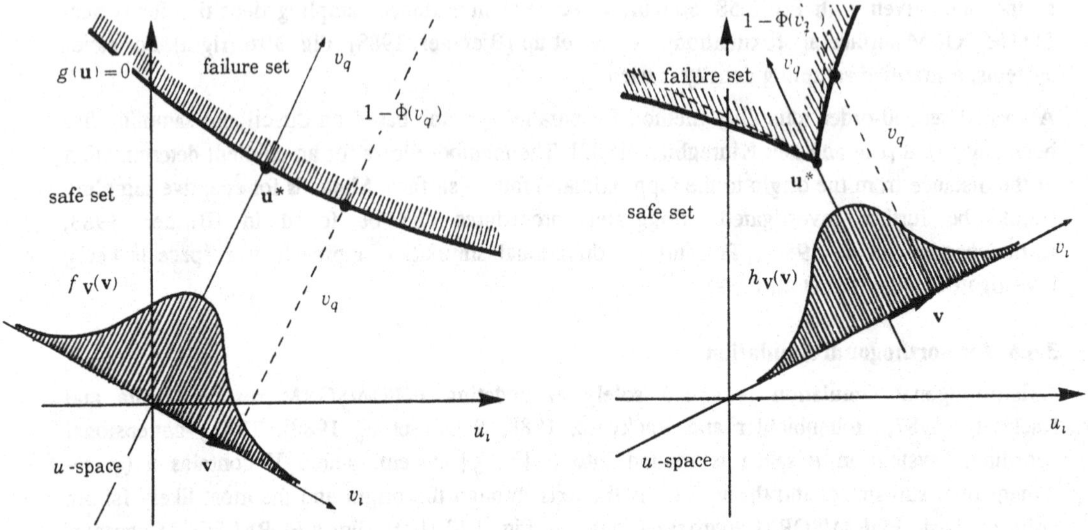

Fig. 3.17. Axis-orthogonal sampling for component (left) and for parallel system (right).

3.6.7 Characteristics of conditional expectation methods

Generality: The methods are based on u-space and require therefore the basic random variables to be continuous. The failure functions should be continuous, whereas there is no restrictions regarding the differentiability of the functions.

Accuracy: For sample size $N \to \infty$, the estimated probability converges to the exact result. For finite sample sizes, uncertainty estimates on the results are provided.

Efficiency: As a general rule, the CPU-time grows approximately linearly with $1/p$ and q for given coefficient of variation on the estimator. Moreover, the enhanced efficiency as compared with indicator-based simulation schemes, generally decreases with q, i.e. the effect of the analytical probability integration vanishes asymptotically for large q. The absolute computation time depends on the time necessary to evaluate the failure functions. For small order probabilities the methods can be expensive in CPU-time.

3.6.8 Sensitivity by simulation

With the analytical conditional simulation methods, parametric sensitivity can be simulated directly by sampling the derivative - with respect to the parameter in question - of the sample values above, [Bjerager and Krenk, 1987; Ditlevsen and Bjerager, 1988]. The methods are therefore also applicable to compute densities on surfaces as well as conditional reliabilities. For sensitivity of mid-range probabilities, the simulation methods may be preferable to the first order results.

For example, in case of a single root $r_i = r(\mathbf{a}_i)$ for the distance to the failure surface, the sample value in uniform directional simulation becomes

$$\frac{dp_i}{d\theta} = \frac{2r_i}{\nabla g(r_i \mathbf{a}_i)^T \mathbf{a}_i} \cdot \frac{d g(r_i \mathbf{a}_i)}{d\theta} \tag{3.64}$$

For indicator-based simulation it is usually suggested that re-sampling for sensitivity analysis [Melchers, 1988] be applied. However, for parametric sensitivity with respect to distribution parameters (mean values, standard deviations, etc.) this is in fact not necessary [Karamchandani et al, 1988]. The sample value can simply be taken as

$$\frac{dp_i}{d\theta} = I[G(\mathbf{x}_i) \le 0] \cdot \frac{d f_{\mathbf{X}}(\mathbf{x}_i;\theta)}{d\theta} \cdot \frac{1}{h_{\mathbf{X}}(\mathbf{x}_i)} \tag{3.65}$$

It is seen, that the derivative of the density with respect to θ is needed only for the points in the failure set, i.e. for $I[G(\mathbf{x}_i) \le 0] = 1$.

Parametric sensitivity with respect to failure function parameters must be estimated by (costly) re-runs. In fact this type of parametric sensitivity is very important due to the numerous and vital applications of conditional reliabilities, and more research to improve the efficiency of such indicator-based MCS is therefore needed.

It is noted, that the optimal sampling density for a parametric sensitivity typically *differs* from the optimal sampling density with respect to reliability computation. Therefore, the choice of sampling densities needs careful consideration.

3.7 RESPONSE SURFACE METHODS

A well-known technique from classical, statistical methods is the response surface method in which a complex (computer) model is approximated by a simple functional relationship between the output quantities and the input (basic) variables, see [Box and Draper, 1987; Iman and Helton, 1985]. Often, linear or quadratic response functions are applied. Adopting the simpler response functions allows an efficient repeated computations, for example as may be needed in simulations or parameter studies.

Let the true relationship between the output Z and the basic variables $\mathbf{X}$ be $Z = G(\mathbf{X})$, where G is defined by the (computer) model. Now, the model function G is approximated by the response function Γ, which, when using standard methods are fitted such that the first two moments of $\Gamma(\mathbf{X})$ equals those of $G(\mathbf{X})$, i.e., $E[\Gamma(\mathbf{X})] \approx E[G(\mathbf{X})]$ and $Var[\Gamma(\mathbf{X})] \approx Var[G(\mathbf{X})]$. Various techniques to determine an appropriate approximation Γ exists, such as those based on experimental design schemes [see Box and Draper, 1987; Veneziano et al., 1982].

In structural reliability studies the concept of response surface methods has been used when approximating costly-to-compute and/or non-differentiable limit state functions. The need for this may arise when the appropriate computer model is complex to realistically model the underlying mechanical problem. Examples of such studies are given e.g. in [Bucher and Bourgund, 1987; Faravelli 1989 and 1990, Holm 1990].

As structural reliability typically is governed by the tail behaviour of the probability distribution of the safety margin $G(\mathbf{X})$ it is apparent that the classical response surface methods based on second moment characteristics may not be sufficient in reliability studies. Ideally, the response surface for reliability purposes should be fitted such that the failure probability remains unchanged, i.e., $P[\Gamma(\mathbf{X}) \le 0] \approx P[G(\mathbf{X}) \le 0]$. Furthermore, it could be required that the most likely failure points are the same for the approximating and original failure surface, or that the (generalized) importance measures as defined by [Hohenbichler and Rackwitz, 1986a] are the same for the two models.

No general scheme has been developed to efficiently established linear and quadratic response surfaces for reliability computations. Adaptive techniques that combine FORM/SORM calculations on the response surface function with point evaluations of the true failure function has been suggested e.g. in [Rackwitz, 1982; Karamchandani, 1987; Holm, 1990]. Further research is needed to establish general and robust response surface methods for high-reliability problems of many basic variables.

3.8 INFORMATION UPDATING (CONDITIONAL RELIABILITY)

One of the promising applications of probabilistic methods is for reliability updating based on improved or new information. For example, information obtained by inspections can be used to update (often increase) the reliability with respect to fatigue crack growth in weldings of offshore structures, and online monitoring of structural performance may give improved information about the dynamic properties of a structure.

The information may be given in one of the following forms:

Direct or *sampling* information about the basic variables can be used to modify or update the probability distributions of these. Standard methods from Bayesian analysis should be used for the updating.

Indirect or *relational* information - for example monitored displacements of a loaded structure - is represented by an event expressed by a function of the basic variables, i.e. $\{H(\mathbf{X}) \le 0\}$, or $\{H(\mathbf{X}) = 0\}$. The indirect information is then accounted for, in the reliability assessment, by considering the conditional reliability

$$p = P[G(\mathbf{X}) \le 0 \mid H(\mathbf{X}) \le 0] = \frac{P[G(\mathbf{X}) \le 0 \cap H(\mathbf{X}) \le 0]}{P[H(\mathbf{X}) \le 0]}$$

$$= \frac{\int_{G(\mathbf{x}) \le 0 \cap H(\mathbf{x}) \le 0} f_{\mathbf{X}}(\mathbf{x}) d\mathbf{x}}{\int_{H(\mathbf{x}) \le 0} f_{\mathbf{X}}(\mathbf{x}) d\mathbf{x}} \tag{3.66}$$

The conditional reliability in Eq. 3.66 can be found as the ratio between the reliability for a parallel system and the reliability for a component. As a special case the distribution of the basic variables can be updated by indirect information simply by using $G(\mathbf{X}) = X_i - x_i$ for different values of x_i.

In the case of information expressed as an equality event margin, the conditional reliability to consider is

$$p = P[G(\mathbf{X}) \le 0 \mid H(\mathbf{X}) = 0] \tag{3.67}$$

$\{H(\mathbf{X}) = 0\}$ is an event of zero probability, and care should be given to specify how the conditional probability is defined. Examples of FORM/SORM computations of such conditional probabilities are given in [Madsen, 1987; Schall and Rackwitz, 1988; Ronold, 1989]. In [Schall and Rackwitz, 1988] it is shown how the surface approximation can take place in the most likely points on the equality constraint in the failure set.

3.8.1 Conditioning in Gaussian reliability problems; proof loading

Consider the basic $R-S$ reliability model in Fig. 3.2. Before the actual loading S (e.g. the life time maximum load) is applied to the structure (bar), the structure is subjected to a proof loading Q. A schematical illustration of this idealized load history is given in Fig. 3.18.

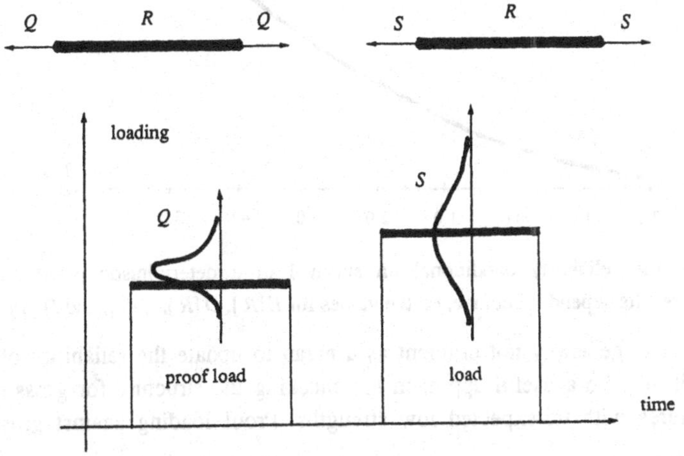

Fig. 3.18. Schematical illustration of structure with resistance R subjected first to the proof load Q and then to the actual load S.

The structure is assumed to survive the proof loading and that no damaged to the structure is caused by this loading, i.e., that the structural resistance R remains unchanged. The event margin describing the proof loading is

$$H = R - S \tag{3.68}$$

and the event of survival is $\{H > 0\}$. The conditional failure probability for the structure subjected to the load S given survival of the proof loading is

$$P[R-S \leq 0 \mid R-Q > 0] = P[M \leq 0 \mid H > 0] = \frac{P[M \leq 0 \cap H > 0]}{P[H > 0]} \tag{3.69}$$

where M is the safety margin $M = R-S$. When all three basic variables are Gaussian, the conditional reliability is easily computed by use of the one- and two-dimensional cumulative normal distribution functions Φ and Φ_2. For example,

$$P[M \leq 0 \mid H > 0] = \frac{\Phi_2(-\beta_M, \beta_H; \rho_{MH})}{\Phi(\beta_H)} \tag{3.70}$$

where β_M and β_H are the elementary reliability indices for the Gaussian safety margin M and event margin H, respectively, and ρ_{MH} is the correlation coefficient.

In Fig. 3.19 results for the conditional reliability index is given by the reliability index $\beta_{M \mid H} = \Phi^{-1}(1-P[M \leq 0 \mid H > 0])$ for a deterministic proof load of size $Q = E[R] + \gamma D[R]$. It is seen that if the proof loading shall have a significant effect on the conditional reliability the proof loading must be rather large as compared to the strength of the structure, i.e., the failure probability under proof loading must be significant.

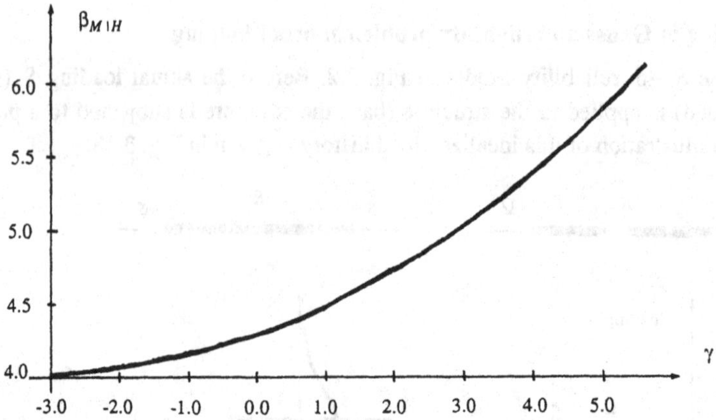

Fig. 3.19. Results on the reliability conditional on survival of a deterministic proof load of size $Q = E[R]+\gamma D[R]$. The results depend, of course, on the values for $E[R]$, $D[R]$, $E[S]$, and $D[S]$.

As seen, proof loading is generally not efficient as a mean to update the reliability of an intact structure. However, it may be a useful approach for checking the structure for gross errors, i.e. events causing structures with un-expected low strengths. Proof loading against gross erros is illustrated in Fig. 3.20.

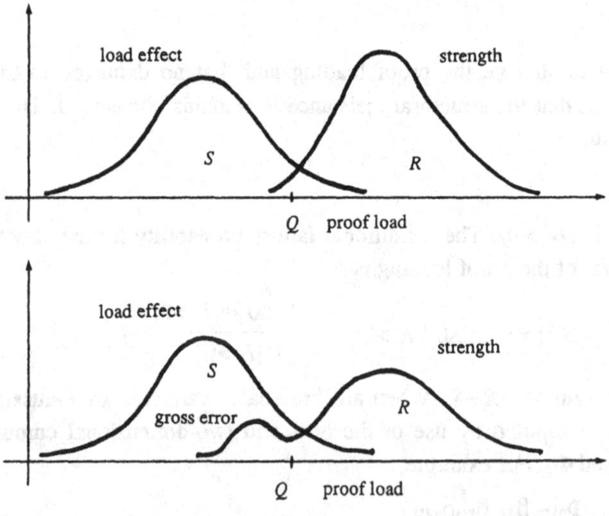

Fig. 3.20. Illustration of proof loading on population of intact structures (top) and of structures with strength degradations due to gross errors as occurring e.g. under construction (bottom).

In the Gaussian case it is also simple to set up the expressing for the conditional reliability when the conditioning event is expressed by an equality event, i.e. H=0. The expression can be derived by utilizing that the conditional safety margin $M\,|H=h$ is Gaussian and then compute the mean and standard deviation for this directly from known formulas from probability theory, i.e.,

$$E[H\,|H=h] = E[M] + \rho_{MH} D[M]\frac{h-E[H]}{D[H]} \qquad (3.71)$$

$$Var[M \mid H=h] = Var[M](1-\rho_{MH}^2) \tag{3.72}$$

whereby the conditional reliability index for $h=0$ is

$$\beta_{M \mid H=0} = \frac{\beta_M - \rho_{MH}\beta_H}{\sqrt{1-\rho_{MH}^2}} \tag{3.73}$$

Alternatively, the conditional failure probability can be derived by using that the probability density $d(h)$ for H is given by

$$d(h) = \frac{d}{dh} P[H \le h] \tag{3.74}$$

and then express the conditional probability by

$$P[M \le 0 \mid H=h] = \frac{\frac{\partial}{\partial h} P[M \le 0 \cap H \le h]_{h=0}}{\frac{\partial}{\partial h} P[H \le h]_{h=0}} \tag{3.75}$$

3.8.2 Time-dependent reliability problems

Reliability updating has been extensively used in problems with degradation of the strength over time. The degration is caused by damage such as fatigue, crack growth, creep, corrosion, erosion and wear. In Fig. 3.21 a monotone degradation process is shown. The damage $D(t)$, say, crack length or reduction of structural dimension due to wear or corrosion, develops over time.

The structure is defined to fail when the damage reaches some critical damage $D_{critical}$. This critical damage could be dependent on a number of other factors, such as static overloading in the case of a limit state for unstable crack crowth. The structure is considered in a reference period T, e.g. the design life time. The damage at the end of this period is $D[T]$ and the failure probability with respect to the reference period is $p_F[T] = P[D(T) > D_{critical}]$ as shown in Fig. 3.21 (top).

Now, if inspections of the damage is carried out through the life time of the structure, the reliability can be updated with respect to this information. Also the inspection result is associated uncertainty for example as described by probability of detection curves (POD-curves) in fatigue crack inspections. The observation introduces a "probabilistic filtering" of the likelihood of the different outcomes of the damage process $D(t)$, and because of this the failure probability given the inspection I can be updated to $P_{F \mid I}[T]$. Also, for example, the distribution of the initial damage $D_{initial}$ may be updated, as shown in Fig. 3.21.

To use the random variable reliability models studied in this paper, the description of the development of the damage should be given by a random variable model. This is the case for commonly applied models for crack growth, see e.g. [Madsen, 1987]. Also, there should be a balance between the realism of the model and the number and extent of observations on which the reliability is to be updated.

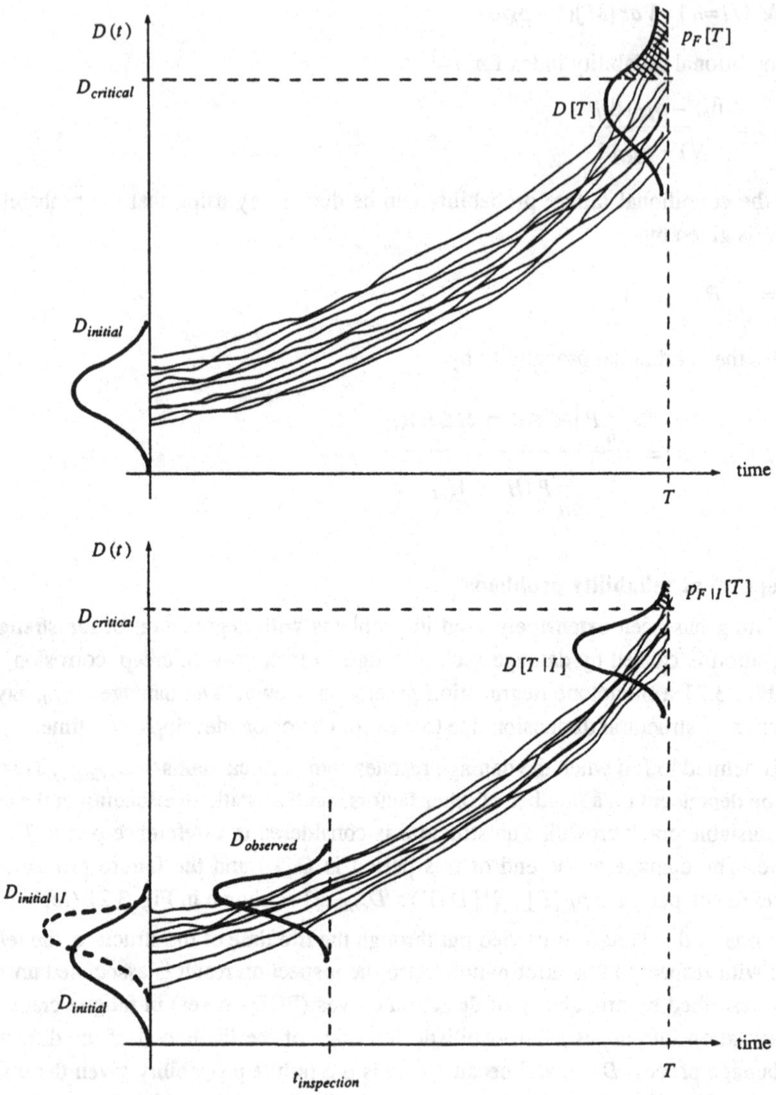

Fig. 3.21. Illustration of reliability problem with a time-dependent damage $D(t)$. The bottom illustration shows one observation in the reference period T.

3.8.3 Life time distribution and the hazard function

In time-dependent reliability problems, the distribution of the life time $F_T(t)$ has interest. The relation between this and the failure probability with respect to a reference period t is

$$F_T(t) = P[T_{life} \leq t] = p_F[t] \tag{3.76}$$

An important concept is the hazard function, also known as the conditional failure rate, which is defined by

$$h(t) = \lim\{dt \to 0\}\, P\,[t \le T_{life} \le t + dt \mid t < T_{life}] = \frac{f_T(t)}{1 - F_T(t)} \tag{3.77}$$

where $f_T(t)$ is the life time density function. It is seen that the hazard function expresses the conditional probability that the structure fails in the next period of dt given that the structure survived until this point in time. The hazard function plays an important role when making decisions regarding the operation of a structure.

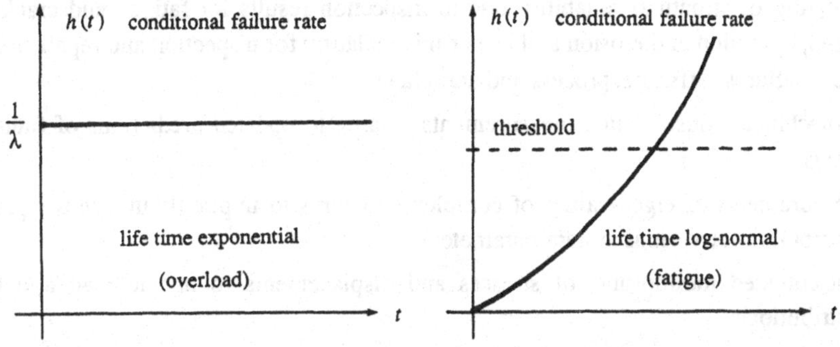

Fig. 3.22. Illustration of hazard functions for two idealized cases of overload failure (left) and fatigue failure (right).

In Fig. 3.22 two typical hazard rates are shown, namely the constant hazard function corresponding to an exponential life time distribution (a no-memory process) and a log-normal life time distribution. The exponential distribution can be taken as a representive distribution for the overload failure of a structure in a long term stationary loading environment and with a constant-over-time strength. In this case the likelihood of failure due to a storm is independent of the age of the structure. The second case can be taken as representative for the hazard function for a structure exposed to damage resulting in strength degradation. This is the case for fatigue-sensitive structures where the likelihood of failure increases throughout the life time of the structure. When the conditional failure rate (the hazard function) exceeds a critical value (the threshold as shown in Fig. 3.22), the structure is not considered safe for continued operation, and it must be futher investigated, repaired or condemned.

3.8.4 Conditional reliability computation

Conditional reliability *computation* as suggested, e.g., by [Wen, 1987] for the case of random process reliability problems has general and wide applications within FORM/SORM probability computations. For example, let (**Y**,**Z**) be two sets of random variables. If the conditional probability

$$p(\mathbf{y}) = P\,[G(\mathbf{Z},\mathbf{Y}) \le 0 \mid \mathbf{Y} = \mathbf{y}] \tag{3.78}$$

is easy to compute, a *nested* FORM/SORM computation of the total probability $P\,[G(\mathbf{Z},\mathbf{Y}) \le 0]$ may be preferable. Such a computation consists of computing by FORM/SORM the probability $P\,[U - \Phi^{-1}(p(\mathbf{Y})) \le 0]$, where U is an auxiliary normalized Gaussian variable and Φ^{-1} is the inverse standard normal distribution function, and during this computation to also compute the conditional probability in Eq. 3.78 by FORM/SORM, [Wen, 1987]. If the probability distribution of the random variable $p(\mathbf{Y})$ is wanted, a safety margin of the form $G(\mathbf{Y})=p(\mathbf{Y})-\theta\leq 0$ can be considered for different θ values. This latter application is found within an analysis of statistical uncertainty, see [Der Kiureghian, 1988] and the references therein.

3.8.5 Practical applications of reliability updating

Probabilistic reliability updating has found many interesting applications. Some of these are:

- updating of structural reliability due to inspection results for fatigue and crack growth, creep, corrosion and erosion and use of this updating for inspection and repair planning of the structures (offshore, process and aeroplane)
- geotechnical consolidation measurements leading to updated predictions of future settlements
- measurements of eigen-values of complex structures to implicitly update the probability distribution for various system parameters
- instrumented monitorings of stresses and displacements to update load and response distributions
- perform history-matching in evaluation of an oil reservoir
- updating of probabilistic financial risk models taking into account observed performance

It is expected that the use of information updating will increase in the future as the computational capabilities improve.

3.9 CONCLUSION

The state-of-the-art of reliability computation methods for structural and mechanical engineering has been outlined, based on a classification of mathematical reliability models into 1) random variable reliability models, 2) random process reliability models, and 3) random field reliability models.

The computational abilities for the different models have been evaluated and discussed. These abilities decrease as the level of the uncertainty description increase. For random variable models the computational ability is in good standing. Many fundamental problems have been solved over the last decade. For higher level models involving e.g. random fields the computational abilities are very restricted. However, the use of random processes and fields in the formulation of discretised models for e.g. FERM is very important.

Besides the computation of the reliability, the possibilities of computing conditional reliabilities, as well as sensitivity measures such as parametric sensitivity, have also been addressed. The available methods for sensitivity computation are less developed than the methods for probability integration. Application of these concepts are very wide and important, and many implementations will be seen in the future.

The two main classes of computation methods for random variable reliability models, the analytical first- and second-order reliability methods, and the Monte Carlo simulation methods, have been described in detail, and the strong and weak points of the methods have been discussed. To this end it is noted that FORM/SORM and the various variance reduction techniques *focus* according to properties of the considered G-function. This implies that the number of relative "expensive" (non-linear etc.) computations in such analyses will be relatively higher than in a direct Monte Carlo simulation, where many analyses are done around the mean values. However, if the aim is to analyze a large system with many and varied criteria, the direct Monte Carlo simulation analyses all criteria simultaneous whereas the focused approaches generally must be run for each criteria separately. In such cases, direct simulation can be superior as regards computational efficiency.

In closing it it mentioned that it is commonly stated that the CPU-power of today's computer grows so rapidly that we maybe need not to invest effort in developing more advanced reliability computation method because soon all analyses can be run efficiently by direct Monte Carlo simulation. However, it should be remembered that there will always be a demand for increasing the realism in the mechanical modeling/analysis and therefore, there will in the future remain a constant challenge for the reliability engineer to develop as efficient reliability computation methods as possible.

REFERENCES

Ang, H.S., and Tang, W.H. (1984): *Probability Concepts in Engineering Planning and Design, Vol. II*, John Wiley and Sons, New York.

Arnbjerg-Nielsen, T., and Bjerager, P. (1988): "Finite Element Reliability Method with Improved Efficiency by Sensitivity Analysis", *Computational Probabilistic Methods*, ASME, AMD-Vol. 93, ed. by Liu, Belytschko, Lawrence and Cruse, 15-25.

Arora, J.S. (1988): "Computational Design Optimization: A Review and Future Directions", presented at the NSF Workshop on Structural System Reliability, University of Colorado, Boulder, Colorado, September 12-14, 1988.

Augusti, G., Baratta, A., and Casciati, F. (1984): *Probabilistic Methods in Structural Engineering*, Chapman and Hall, London.

Barlow, R.E. and Proschan, F. (1981): *Statistical Theory of Reliability and Life Testing*, To Begin With, Silver Springs, Maryland,

Bjerager, P., and Krenk, S. (1987): "Sensitivity Measures in Structural Reliability Analysis", *Proc. of 1st IFIP Working Conference on Reliability and Optimization on Structural Systems*, ed. by P. Thoft-Christensen, Springer Verlag, 459-470.

Bjerager, P. (1988): "Probability Integration by Directional Simulation", *Journal of Engineering Mechanics*, ASCE, 114(8), 1285-1302.

Bjerager, P., Winterstein, S.R., and Cornell, C.A. (1988a): "Outcrossing Rates by Point Crossing Method", *Probabilistic Engineering Mechanics*, ASCE, ed. by P.D. Spanos, 533-536.

Bjerager, P., Loseth, R., Winterstein, S., and Cornell, C.A. (1988b): "Reliability Method for Marine Structures under Multiple Environmental Load Processes", *Proceedings of 5th International Conference on the Behaviour of Offshore Structures*, Vol. 3, Trondheim, Norway, 1239-1253.

Bjerager, P. (1989): "Plastic Systems Reliability by LP and FORM", *Computers and Structures*, 31(2), 187-196.

Bjerager, P. (1990): "Short Communication on Omission Sensitivity Factors", *Structural Safety*, &, 77-79.

Bjerager, P., and Krenk, S. (1989): "Parametric Sensitivity in First Order Reliability Analysis", *Journal of Engineering Mechanics*, ASCE, 115(7).

Box, G.E.P, and Draper, N.R. (1987): *Empirical Model-Building and Response Surfaces*, John Wiley and Sons, New York.

Breitung, K. (1984): "Asymptotic Approximation for Multinormal Integrals", *Journal of Engineering Mechanics*, ASCE, 110(3), 357-366.

Bucher, C.G., and Bourgund, U., (1987): Efficient Use of Response Surface Methods, Report 87-9, Institute for Engineering Mechanics, University of Innsbruck, Austria.

Bucher, C.G. (1988): "Adaptive Sampling: An Iterative Fast Monte-Carlo Procedure", *Structural Safety*, 5(2), 119-126.

Cornell, C.A. (1969): "A Probability-Based Structural Code", *Journal of the American Concrete Institute*, 66(12), 974-985.

Cruse, T.A., Wu, Y.-T., Dias, S., and Rajagopal, K.R. (1988): "Probabilistic Structural Analysis Methods and Applications", *Computers and Structures*, 30(1/2), 163-170.

Deak, I. (1980): "Three Digit Accurate Multiple Normal Probabilities", *Numerische Mathematik*, 35, 369-380.

Der Kiureghian, A. (1985): "Finite Element Based Reliability Analysis of Frame Structures", *Structural Safety and Reliability*, ed. by Konishi, Ang and Shinozuka, ICOSSAR, Kobe, Japan, Vol. I., 395-404.

Der Kiureghian, A., and Liu, P.-L. (1986): "Structural Reliability under Incomplete Probability Information", *Journal of Engineering Mechanics*, ASCE, 112(1), 85-104.

Der Kiureghian, A., Lin, H.-Z., and Hwang, S.J. (1987): "Second-Order Reliability Approximations". *Journal of Engineering Mechanics*, ASCE, 113(8), 1208-1225.

Der Kiureghian, A., and Ke, J.-B. (1987): "The Stochastic Finite Element Method in Structural Reliability", *Probabilistic Engineering Mechanics*, 3(2), 83-91.

Der Kiureghian, A. (1988): "Measures of Structural Safety under Imperfect States of Knowledge", accepted for *Journal of Structural Engineering*, ASCE.

Der Kiureghian, A., and De Stefano, M. (1990): "An Efficient Algorithm for Second-Order Reliability Analysis", Report No. UCB/SEMM-90/20, Department of Civil Engineering, University of California at Berkeley, Berkeley, California.

Ditlevsen, O. (1973): "Structural Reliability and the Invariance Problem". Research Report No. 22, Solid Mechanics Division, University of Waterloo, Ontario, Canada.

Ditlevsen, O. (1981): *Uncertainty Modeling with Applications to Multidimensional Civil Engineering Systems*, McGraw-Hill Inc, New York.

Ditlevsen, O. (1983): "Gaussian Outcrossings from Safe Convex Polyhedrons". *Journal of Engineering Mechanics*, ASCE, 109(1), 127-148.

Ditlevsen, O., and Bjerager, P. (1986): "Methods of Structural Systems Reliability", *Structural Safety*, 3, 195-229.

Ditlevsen, O., Bjerager, P., Olesen, R., and Hasofer, A.M. (1988): "Directional Simulation in Gaussian Processes", *Probabilistic Engineering Mechanics*, 3(4), 207-217.

Ditlevsen, O., Melchers, R., and Gluver, H. (1989): "General Probability Integration by Directional Simulation", Manuscript, Department of Structural Engineering, Technical University of Denmark.

Ditlevsen, O., and Madsen, H.O. (1990): "Structural reliability" (in Danish), SBI-Report 211, The Danish Building Research Institute, Hoersholm, Denmark.

Faber, M, and Rackwitz, R. (1988): "Excursion Probabilities of Non-Homogeneous Gaussian Scalar Fields based on Maxima Considerations", *Proc. of 2nd IFIP Working Conference on Reliability and Optimization on Structural Systems*, ed. by P. Thoft-Christensen, Springer Verlag (in press).

Faravelli, L. (1989): "Response-Surface Approach for Reliability Analysis", *Journal of Engineering Mechanics*, ASCE, 115(12), 2763-2781.

Faravelli, L., and Bigi, D. (1990): "Stochastic Finite Elements for Crash Problems", *Structural Safety*, 8, 113-130.

Fiessler, B., Neumann, H.-J., and Rackwitz, R. (1979): "Quadratic Limit States in Structural Reliability", *Journal of Engineering Mechanics Division*, ASCE, 105, 661-676.

Fujita, M., and Rackwitz, R. (1988): "Updating First and Second Order Reliability Estimates by Importance Sampling", *Structural Engineering and Earthquake Engineering*, JSCE, 5 (1), 31s-37s.

Gollwitzer, S., and Rackwitz, R. (1988): "An Efficient Numerical Solution to the Multinormal Integral", *Probabilistic Engineering Mechanics*, 3(2), 98-101.

Hagen, O., and Tvedt, L. (1990): "Vector Process Outcrossing as a Parallel System Sensitivity Measure". Submitted to *Journal of Engineering Mechanics*, ASCE, October 1990.

Hagen, O. (1991): "Conditional and Joint Crossing of Stochastic Processes". Submitted to *Journal of Engineering Mechanics*, ASCE, January 1991.

Hagen, O., and Tvedt, L. (1991): "Parallel System Approach for Vector Outcrossings". To appear in *Proceedings of the Tenth International Conference on Offshore Mechanics and Arctic Engineering, Stavanger, Norway, 1991*.

Harbitz, A. (1983): "Efficient and Accurate Probability of Failure Calculation by use of the Importance Sampling Technique". *Proc. of ICASP4*, Firenze, Italy, 825-836.

Hasofer, A.M. and Lind, N.C. (1974): "Exact and Invariant Second-Moment Code Format." *Journal of the Engineering Mechanics Division*, ASCE, 100(EM1), 111-121.

Hasofer, A.M., Ditlevsen, O., and Olesen, R. (1987): "Vector Outcrossing Probabilities by Monte Carlo", Report 349, Danish Center for Applied Mathematics and Mechanics, Technical University of Denmark.

Hohenbichler, M., and Rackwitz, R. (1981): "Non-normal Dependent Vectors in Structural Reliability", *Journal of Engineering Mechanics Division*, ASCE, 107, 12127-1238.

Hohenbichler, M. (1984): "An Asymptotic Formula for the Probability of Intersections", Berichte zur Zuverlassigkeitstheorie der Bauwerke, SFB 96, Technical University of Munich, Heft 69, 21-48.

Hohenbichler, M., and Rackwitz, R. (1986a): "Sensitivity and Importance Measures in Structural Reliability", *Civil Engineering Systems*, 3(4), 203-209.

Hohenbichler, M., and Rackwitz, R. (1986b): "Asymptotic Crossing Rate of Gaussian Vector Processes into Intersections of Failure Domains", *Probabilistic Engineering Mechanics*, I(3), 177-179.

Hohenbichler, M., Gollwitzer, S., Kruse, W., and Rackwitz, R.: "New Light on First- and Second-Order Reliability Methods", *Structural Safety*, 4, 267-284.

Hohenbichler, M., and Rackwitz, R. (1988): "Improvement of Second-Order Reliability Estimates by Importance Sampling", *Journal of Engineering Mechanics*, ASCE, 114(12), 2195-2199.

Holm, C.A. (1990): Reliability Analysis of Structural Systems Using Nonlinear Finite Element Methods, Division of Structural Mechanics, The Norwegian Institute of Technology, University of Trondheim, Norway.

Igusa, T., and Der Kiureghian, A. (1988): "Response of Uncertain Systems to Stochastic Excitation", *Journal of Engineering Mechanics*, ASCE, 114(5).

Iman, R.L., and Helton, J.C. (1985): "A Comparison of Uncertainty and Sensitivity Analysis Techniques for Computer Models". Report NUREG/CR-3904, Sandia National Laboratories, Albuquerque, New Mexico.

Karamchandani, A. (1987): "Structural System Reliability Analysis Methods", Report 83, John A. Blume Earthquake Engineering Center, Stanford University.

Karamchandani, A., Bjerager, P., and Cornell, C.A. (1988): "Methods to Estimate Parametric Sensitivity in Structural Reliability Analysis", *Probabilistic Engineering Mechanics*, ASCE, ed. by P.D. Spanos, 86-89.

Karamchandani, A., Bjerager, P., and Cornell, C.A. (1989): "Adaptive Importance Sampling", *Proceedings*, 5th International Conference on Structural Safety and Reliability, San Francisco, California, August 7-11, 1989.

Karamchandani, A., Dalane, J.I., and Bjerager, P. (1991): "A Systems Approach to Fatigue of Structures". Accepted for publication in *Journal of Engineering Mechanics*, ASCE.

Lin, H.-Z., and Der Kiureghian, A. (1987): "Second-Order System Reliability using Directional Simulation", *Reliability and Risk Analysis in Civil Engineering 2, ICASP5*, ed. by N.C. Lind, University of Waterloo, Ontario, Canada, 930-936.

Liu, P.-L., and Der Kiureghian, A. (1988a): "Reliability of Geometrically Nonlinear Structures", *Probabilistic Engineering Mechanics*, ASCE, ed. by P.D. Spanos, 164-167.

Liu, P.-L., and Der Kiureghian, A. (1988b): "Optimization Algorithms for Structural Reliability" *Computational Probabilistic Methods*, ASME, AMD-Vol. 93, ed. by Liu, Belytschko, Lawrence and Cruse, 185-196.

Liu, W.K., Belytschko, T., and Mani, A. (1986): "Random Field Finite Elements", *Numerical Methods in Engineering*, 23, 1831-1845.

Liu, W.K., Mani, A., and Belytschko, T. (1987): "Finite Element Methods in Probabilistic Mechanics", *Probabilistic Engineering Mechanics*, 2(4), 201-213.

Liu, W.K., Besterfield, G.H., and Belytschko, T. (1988): "Variational Approach to Probabilistic Finite Elements", *Journal of Engineering Mechanics*, ASCE, 114(12), 2115-2133.

Madsen, H.O., Krenk, S., and Lind, N.C. (1986): *Methods of Structural Safety*, Prentice-Hall, Inc., Englewood Cliffs, N.J.

Madsen, H.O. (1987): "Model Updating in Reliability Theory". *Reliability and Risk Analysis in Civil Engineering 1, ICASP5*, ed. by N.C. Lind, University of Waterloo, Ontario, Canada, 564-577.

Madsen, H.O., and Moghtaderi-Zadeh, M. (1987): "Reliability of Plates under Combined Loading", *Proceedings*, Marine Structural Reliability Symposium, SNAME, Arlington, Virginia, 185-191.

Madsen, H.O., and Tvedt, L. (1988): "Efficient Methods in Time Dependent Reliability", *Probabilistic Engineering Mechanics*, ASCE, ed. by P.D. Spanos, 432-435.

Madsen, H.O. (1988): "Omission Sensitivity Factors", *Structural Safety*, 5, 35-45.

Madsen, H.O., and Tvedt, L. (1990): "Methods for Time-Dependent Reliability and Sensitivity Analysis", *Journal of Engineering Mechanics*, ASCE, 116(10), 2118-2135.

Madsen, H.O. (1990): "Sensitivity Factors for Parallel Systems", Submitted to *Journal of Engineering Mechanics*.

McKay, M.D., Beckman, R.J., and Conover, W.J. (1979): "A Comparison of Three Methods for Selecting Values of Input Variables in the Analysis of Output from a Computer Code", *Technometrics*, 21(2).

Melchers, R.E. (1984): "Efficient Monte-Carlo Probability Integration", Report No. 7/1984, Civil Engineering Research Reports, Monash University, Victoria, Australia.

Melchers, R.E. (1987): *Structural Reliability, Analysis and Prediction*, Ellis Horwood Series in Civil Engineering, Halsted Press, England.

Paloheimo, E., and Hannus, M. (1974): "Structural Design Based on Weighted Fractiles", *Journal of the Structural Division*, ASCE, 100(ST7), 1367-1378.

Plantec, J.-Y., and Rackwitz, R. (1989): "Structural Reliability under Non-Stationary Gaussian Vector Process Loads", *Proceedings of the Eighth International Conference on Offshore Mechanics and Arctic Engineering, The Hague, The Netherlands, 1989*.

Powell, M.J.D.: "VMCWD: A FORTRAN Subroutine for Constrained Optimization", Report DAMTP 1982/NA4, Cambridge University, England, 1982.

PROBAN (1989): The PROBabilistic ANalysis Program, Version 2. Theory Manual: VR Report 89-2022, Users Manual: VR Report 89-2024, Example Manual: VR Report 89-2025, Distribution Manual: VR Report 89-2026, and Command Reference: VR Report 89-2027, A.S Veritas Research, Norway.

Rackwitz, R. (1982): "Response Surfaces in Structural Reliability", Report 67/1982, Institut fur Bauingenieurwesen III, Technical University of Munich, West Germany.

Rackwitz, R. (1988): "Asymptotic Integrals in Reliability Theory" (in German), Aus Unseren Forschungsarbeiten VI, Lehrstuhl fur Massivbau, Technical University of Munich, West Germany, 175-181.

Rice, S.O. (1944). "Mathematical Analysis of Random Noise", *Bell System Technical Journal*, 23(282), 24(46).

Ronold, K.O. (1989): "Probabilistic Consolidation Analysis with Model Updating", *Journal of Geotechnical Engineering*, ASCE, 115(2).

Rubinstein, R.Y. (1981): *Simulation and the Monte Carlo Method.* J. Wiley & Sons, New York.

Schall, G., Gollwitzer, S., and Rackwitz, R. (1988): "Integration of Multinormal Densities on Surfaces", *Proc. of 2nd IFIP Working Conference on Reliability and Optimization on Structural Systems*, ed. by P. Thoft-Christensen, Springer Verlag (in press).

Schittkowski, K. (1986): "NLPQL: A Fortran Subroutine Solving Constrained Nonlinear Programming Problems", *Annals of Operations Research*, 5, 485-500.

Shinozuka, M. (1964): "Probability of Failure under Random Loading", *Journal of Engineering Mechanics*, ASCE, 90(5), 147-171.

Shinozuka, M. (1983): "Basic Analysis of Structural Safety". *Journal of Structural Engineering*, ASCE, 109(3), 721-740.

Shinozuka, M. (1987): "Basic Issues in Stochastic Finite Element Analysis", *Reliability and Risk Analysis in Civil Engineering 1, ICASP5*, ed. by N.C. Lind, University of Waterloo, Ontario, Canada, 506-519.

Thoft-Christensen, P., and Murotsu, Y. (1986): *Application of Structural Systems Reliability Theory*, Springer Verlag, West Germany.

Tvedt, L. (1988): "Second Order Reliability by an Exact Integral", *Proc. of 2nd IFIP Working Conference on Reliability and Optimization on Structural Systems*, ed. by P. Thoft-Christensen, Springer Verlag (in press).

Tvedt, L. (1990): "Distribution of Quadratic Forms in the Normal Space - Application to Structural Reliability", *Journal of Engineering Mechanics*, ASCE, 116(6), 1183-1197.

Vanmarcke, E. (1984): *Random Fields, Analysis and Synthesis*, The MIT Press, Massachusetts Institute of Technology, Cambridge, Massachusetts.

Vanmarcke, E., Shinozuka, M., Nakagiri, S., Schueller, G.I., and Grigoriu, M. (1986): "Random Fields and Stochastic Finite Elements", *Structural Safety*, Vol. 3.

Veneziano, D., Grigoriu, M., and Cornell, C.A. (1977): "Vector-Process Models for System Reliability", *Journal of Engineering Mechanics Division*, ASCE, 103(EM3), 441-460.

Veneziano, D., Casciati, F., and Faravelli, L. (1983): "Method of Seismic Fragility for Complicated Systems", *Proceedings of 2nd Committee on the Safety of Nuclear Installations (CSNI): Specialist Meeting on Probabilistic Methods in Seismic Risk Assessment for NPP*, Livermore, California.

Wen, Y.K. (1987): "Approximate Methods for Nonlinear Time-Variant Reliability Analysis", *Journal of Engineering Mechanics*, ASCE, 113(12), 1826-1839.

Wen, Y.K., and Chen, H.-C. (1987): "On Fast Integration for Time Variant Structural Reliability", *Probabilistic Engineering Mechanics*, 2(3), 156-162.

Winterstein, S.R., and Cornell, C.A. (1984): "Load Combination and Clustering Effects", *Journal of Structural Engineering*, ASCE, 110, 2690-2708.

Winterstein, S.R., and Bjerager, P. (1987): "The Use of Higher Moments in Reliability Estimation", *Reliability and Risk Analysis in Civil Engineering 2, ICASP5*, ed. by N.C. Lind, University of Waterloo, Ontario, Canada, 1027-1036.

Winterstein, S.R. (1988): "Nonlinear Vibration Models for Extremes and Fatigue", *Journal of Engineering Mechanics*, ASCE, 114(10), 1772-1790.

Wu, Y.-T., and Wirsching, P.H. (1987): "A New Algorithm for Structural Reliability Estimation", *Journal of Engineering Mechanics*, ASCE, 113, 1319-1336.

Wu, Y.-T., Burnside, O.H., and Cruse, T.A. (1989): "Probabilistic Methods for Structural Response Analysis", *Computational Mechanics of Probabilistic and Reliability Analysis*, ed. by Liu and Belytschko, Elme Press International, Lausanne, Switzerland.

Chapter 4

ENGINEERING, OPERATIONAL, ECONOMIC, AND LEGAL ASPECTS OF THE RELIABILITY ASSURANCE

M. Tichy
Czech Technical University, Praha, Czechoslovakia

Abstract

Four principal groups of aspects govern the reliability assurance process, viz. engineering aspects, operational aspects, economic aspects, and legal aspects. All these groups mutually interact. Usually attention is paid only to the first two groups, but the importance of economic and legal factors in the reliability assessment of a constructed facility is not negligible. - Many partial factors affect the reliability assurance process: theoretical and empirical knowledge, experience; codes, reliability requirements; qualification of personnel, quality assurance and control; target life and target failure probability; economic climate, economic recession and expansion periods, inflation; system of legal documents, liabilities, insurance companies, and other factors.

4.1 INTRODUCTION

The assurance of the reliability of a constructed facility is a continuous decision process which begins with the first intention of building the facility, for a defined purpose, in a defined space and time, and in an assumed environment. This decision process (the reliability assurance process, RAP) ends on the dismantling, or demolition, of the respective facility. Four principal groups of aspects govern RAP and the decisions taken during its development:

- engineering aspects,
- operational aspects,
- legal aspects.

Each group can be discussed and investigated separately but their effects upon the reliability of a constructed facility are mutually dependent. This obvious fact is very often neglected during the education of civil engineers because in university courses on reliability, where they exist, emphasis is usually given to the engineering aspects; the other three groups remain in the shadow of the first one. This is particularly true with respect to the legal aspects group where, on the whole, engineering education is usually surprisingly poor. Unfortunately, also, the mutual dependence of the four groups is, as a rule, not recognized in many reliability monographs.

Let us discuss the four foregoing groups in more detail.

4.2 ENGINEERING ASPECTS

Houses, bridges, dams and roads have been built since the existence of mankind. As a result, beams and cantilevers, columns, arches, retaining walls, or just compact soil are typical systems which constantly appear in all constructed facilities, throughout the ages. We are well acquainted with them and we use them without any trouble, as though they were the elements of some super-Lego set. Conversely, any complicated structural system, such as a space frame, space truss, can be reduced, with more or less effort, to a simply analysable system. This is not a general rule, of course, but it may be reckoned that, historically, the engineering profession always thought to avoid systems which could not be easily simplified. The computer age has brought completely new concepts into the engineer's thinking but the drive towards simplification still prevails in the structural design philosophy.

These observations become more graphic when civil and mechanical engineering systems are compared. Simple machines and tools have also been known for millenia, but they are very far removed from complicated mechanical systems, such as cars, aircrafts, spaceships, or some highly sophisticated computer controlled robots. The design of such systems cannot be based simply on the behaviour of single wheels, levers, wedges, or helices.

The group of engineering factors consists, in the first place, of theoretical and empirical knowledge which both have prominent positions in RAP. Theoretical knowledge is either general, and as such it is covered by university curricula, or particular. The latter is not currently being taught in engineering courses and, consequently, must be gained from specialized literature, by individual consultations, etc. It happens that such knowledge, when needed in a specific practical case, is not easily available or does not exist at all, and therefore, theoretical or experimental research has to be carried out. Theoretical knowledge is often overemphasized in the engineering education as a result of the traditionally academic approach to training programmes. However, demand for practically minded and design experienced teachers has grown in recent years (see Taoka [1]).

The engineer's empirical knowledge comes from two sources: general experience, obtained as a continuous heritage of the past; it is usually conveyed to engineering undergraduates, during their learning period, by means of lectures and tutorials, and to practising engineers, that is, designers, site engineers, etc., by means of books, papers, codes (we shall discuss this particular area of experience transmittance later), and by participation in seminars, conferences, etc. The other source is personal experience resulting from an individual's activities during a professional career. Bad experience, as may be guessed, is about five times more useful than good experience. Experience in reliability assurance is extremely important and it is always a gross mistake to underestimate its value.

As soon as sufficient information on a particular phenomenon becomes available, empirical knowledge treating such phenomenon can be substituted by theoretical knowledge. For instance, it is generally known that in Central Europe the depth of the foundation level should never be less than 1.5 m, to avoid the unfavourable effects of frozen soil on the foundation properties. We are now able to calculate exactly the value of the safe depth, by means of sophisticated soil theories, weather data, statistical and probability-based analysis, etc. Obviously some other value of the minimum depth might be reached by these means, say 1.45 m or 1.57 m. However, nobody would perform such calculations, or accept the new numbers, since experience on this particular point is very strong, and more powerful than any exact study. Moreover, any sophistication in such a case would be a waste of time and money because the results would not justify the research expense, in economic terms.

In the discussion of engineering knowledge <u>tradition</u> must not be omitted. This is a particular branch of experience, fossilized it may be said, which can sometimes have a retrograde effect on RAP. Many cases are known where sticking to tradition has resulted in bad solutions, diminishing the reliability of the system. However, on the other hand, good workmanship is always based on tradition!

4.2.1 <u>Codes</u>

In many cases the theoretical and empirical understanding of a certain phenomenon is not uniform among engineers and engineering experts. Where the respective non-uniformity can affect RAP in an adverse manner, the diversified knowledge, theoretical or empirical, must be unified by means of codes, standards or other similar regulations. Thus, for instance, a particular calculation model for the deflection of reinforced concrete beams, reliability factors, target failure probabilities, material testing procedures, load test evaluation rules, construction methods, and other processes are subjected to codified unification.

Law experts say that codes are legal documents "sui generis", that is, of specific nature, because they do not treat social relationships (relationships between individuals and/or bodies), but <u>relationships of men to the natural and engineering phenomena</u>, and <u>ways of manipulating such phenomena in technology</u>. Codes form a part of the legal system of any developed country, and consequently the actual status of the codes depends very much on the intrinsic nature of the legal system; a large variety of concepts exists. In many countries, <u>design codes</u> are <u>mandatory documents</u>, and the designer and contractor are obliged to follow them, while in other countries codes are <u>optional</u>. The difference in approach is more or less formal since optional codes are being used in the same manner as the mandatory ones. As a rule, insurance companies insist on the use of a design code, even when it is of an optional character.

The influence of the optional and economical factors is now being systematically subjected to harmonization procedures among groups of countries with the aim of achieving a uniform, consistent system. This harmonization is being driven by purely economical needs; by no means is it conditioned by the nobel wishes of scientists who, as individuals, are often not completely satisfied with some of its results and consequences. A very typical and most recent example of a successful code harmonization is the Eurocode system of structural design codes worked out by the members of the European Community. The Eurocodes have not yet been completed but it is now certain that they will form a viable code system.

Two principal aims of any design code, including the action codes, can easily be discerned:

- harmonization of calculation models,
- fixing of reliability levels.

The harmonization of calculation models is important from various points of view. Firstly, it is needed to support the compatibility of different design solutions. Secondly, it further helps the designer to avoid difficulties in selecting the respective analysis procedure. Thirdly, it reduces, in a way, the level of designer's responsibilities. New ideas reach the designer frequently just by the intermediation of a code. It is sometimes necessary to specify, for a particular problem, one single calculation model, but in some other problems it may be desirable to advise the designer that he/she is free to use any model that he/she may think suitable (and, of course, logical) for his/her problem. It is noted here that decisions of bodies, e.g. code committees, or individuals govern the choice of the calculation models.

The reliability of structures designed according to a certain design code is described by a system of proportioning requirements which contain the proportioning parameters: reliability coefficients, action combination coefficients, etc. Again, their values are established by a group or through individual decisions. It has to be mentioned that the values of the proportioning parameters depend partly upon the properties of the respective calculation model. Whenever a calculation model is not based on a clear physical description of the phenomenon under consideration and consequently, simplifications, or empirical constants or functions are entered into that model, this fact must necessarily be reflected by the proportioning parameters. There are cases where the dependence of the proportioning parameters upon the calculation model is very strong; this occurs particularly in non-linear problems of analysis (such as instability of axially loaded members, deflection of non-homogeneous beams, etc.). Then, when the calculation model is changed for some reason, e.g. from a biased model to an unbiased one, the model-dependent proportioning parameters have also to be adjusted in order to achieve the same reliability level that pertained the changes.

4.2.2 Reliability requirements

The topic of the reliability requirements has already been discussed by the Author, elsewhere (see Tichy [2]). Let us introduce here only its main issues. - It can be ascertained that, at the most general level, two principal phenomena are to be dealt with in the design of structures, viz. the attack, A, and the barrier, B, resisting the attack. The meaning of these phenomena differs according to the type of problem under investigation. For instance, the attack may be expressed in terms of actions acting on a structure, or in terms of the bending moment in a cross-section produced by the actions, or in terms of the mid-span deflection of a beam, etc., and analogously the barrier may be represented by the ultimate strength of the structure, the ultimate moment of the cross-section, the limiting deflection, etc.

The attack and the barrier are phenomena that must exist simultaneously, otherwise any design would be meaningless. They obviously form the base of mathematically written requirements, necessary for the proportioning of structures, members, cross-sections, or even materials; thus they may be referred to as formative phenomena. It is important to notice that, in general, the attack can be mathematically described by a vectorial function, which is often reduced to a vector, or, further to a scalar variable. Similarly, the barrier may be represented by a vectorial function or by a scalar variable; no example of a barrier expressed by a vector has yet been found. The attack and the barrier may be mutually dependent.

By merging the two formative phenomena a higher level is reached: then a comprehensive phenomenon, the reliability margin, appears in the investigation. There is no need to elaborate the solution procedure here, for it is fairly well known. Considering, for simplicity, a scalar case, the reliability margin is

$$Z = A - B$$

In the case of multi-component phenomena the reliability margin is defined by the minimum distance between attack and the barrier; thus it remains a scalar variable, though with somewhat curious dimensions.

The attack and the barrier are each described by a set of elementary variables which, in general, are random and can be modelled by respective probability distributions. Now, when the reliability of a certain "Structure-Action-Environment" system must be assured, by design, two reliability requirements relating to the reliability margin, Z, must be complied with:

(1) a requirement concerned with the physical nature of the reliability margin, the so-called physical reliability requirement,

(2) a requirement taking into account the random nature of Z; it can have two different forms:

(2a) a probability-based reliability requirement which relates the probability of adverse realizations of Z, the failure probability $P_f = \Pr(Z < 0)$, to a specified value, the target failure probability, P_{ft} ;

(2b) a statistical reliability requirement which relates the statistical parameters of Z (e.g., mean, μ_Z , standard deviation, σ_Z , coefficient of skewness, α_Z) to their respective target values.

Because Z is, as a rule, time-dependent, time must always be respected, in any reliability requirement. We may thus rewrite the foregoing points into the following formulas:

$$\forall t \in T_{ref} : Z \geq 0 \tag{4.1}$$

$$\forall t \in T_{ref} : P_f \leq P_{ft} \tag{4.2a}$$

$$\forall t \in T_{ref} : \mu_Z \geq \mu_{Zt}, \quad \sigma_Z \leq \sigma_{Zt} \tag{4.2b}$$

where t = a point in time, T_{ref} = the reference period during which the reliability requirement is expected to be fulfilled.

Since zero, 0, and the target failure probability, P_{ft} , are non-dimensional, requirements (4.1) and (4.2) can easily be codified to meet the exigencies of generality. This does not hold true, however, with requirement (4.2b). It is hardly possible, for structures, members and cross-sections to give target values for the respective statistical parameters. This can only be done for some elementary variables (material properties, e.g.). For this reason requirement (4.2b) is currently substituted by

$$\forall t \in T_{ref} : \mu_Z/\sigma_Z \geq \beta_{Zt}$$

where β_{Zt} = the target reliability index. It is now generally known that μ_Z/σ_Z depends upon the mathematical formulation of the reliability margin, and that the Z-invariant Hasofer and Lind reliability index, β^{HL} , is a quantity subjectable to codification within certain confines (for fuller information on β^{HL} see, e.g., Madsen, Krenk, and Lind [3]). Consequently, the statistical reliability requirement becomes

$$\forall t \in T_{ref} : \beta^{HL} \geq \beta_t^{HL}$$

Certain relationships between β^{HL} and P_f can be identified if some local approximations to the calculation model are accepted. This problem will not be discussed here; it has been widely covered in many papers and monographs.

The designers, or other individuals participating in RAP, do not appreciate the high statistical sophistication and therefore only proportioning requirements are specified in the design codes. Similarly, multi-component evaluation requirements are given in codes for quality control testing. They have to be derived from requirements type (4.1) combined with either type (4.2a), or type (4.2c), taking into account the behaviour of the elementary variables. Details of such procedures cannot be given here, due to lack of space. They are not, in any event, important for subsequent discussions in this paper.

4.2.3 Code revisions

Constant and continuous developments in the construction industry, efforts to save materials, energy and labour, and also the need to introduce new structural systems lead to a demand for periodical revisions of codes. The aim of a revision procedure is always to improve the actual code statements, to make them more general, or more exact, and to add clauses covering new knowledge gained during the period elapsed from the last issue of the code. If the new knowledge cannot be implemented without substantial changes to the design method then the system of reliability requirements, including various factors, definitions of input variables, etc., called design format, is also updated during the revisions.

Nobody enjoys extensive code revisions. Any change to the design format is usually a disagreeable intervention into the design concepts and its impact on practical design may have wide economical effects. Some structures, designed according to the old code, need more material when designed according to a new code, or conversely. It is, however, not in the interest of society to increase, through acceptance of updated design methods, consumption of materials and energies in general; local adverse deviations from this rule, must be balanced by savings in other areas covered by the code, or code system.

Whenever a design code is changed, many side effects, some very important, must be expected. Any single change to the proportioning parameters and reliability requirements affects the evaluation parameters, and consequently the testing codes must be thoroughly revised, and a new evaluation format has to be formulated. Here is a large problem area, which has not yet been fully investigated; research is being currently directed at these problems, however.

It is usually not sufficient to simply compare the results obtained by means of the old and new code, and to adjust the proportioning parameters. The code calibration must be made more sophisticated, to match extensions to the code system. In a scientific calibration, defined classes of structures are subjected to investigation as stochastic entities, and the respective proportioning parameters are determined by optimization; this would ensure that structures belonging to the respective class are reliable and economical. At a first step the principle is usually followed that the average result of the design should not be substantially changed by the new code format. This, of course, means that some structures must be needlessly oversized by the new code, and therefore, at the second step, parameters are further adjusted, using past experience.

4.2.4 Code systems

At the present time, every developed country benefits from a system of codes which is concerned with the engineering component of RAP. The legally binding detailed features of such code system are not comparable, from country to country, but the general outlines of the systems are almost identical. As already mentioned above, the following principal activity sections are dealt with by code systems:

- design,
- execution and workmanship,
- testing and quality control, and
- maintenance and use.

The first two sections are usually given prominence, whilst the remaining sections, particularly the last one, are neglected in many code systems. However, all four sections are equally important in RAP and all influence each other, up to a certain degree. Therefore, any revision of a code in one section must be projected into the remaining sections. Such a fact is only rarely recognized during revisions. This brings subsequent difficulties to code makers, and to code users as well.

It should be remarked that code systems and also the individual codes are, in fact, reliability systems themselves! When they are considered from this angle all the main features of regular reliability systems in a particular code, or code system, become apparent: components (= individual codes), burn-in period (= the period after first publication and before first revision), useful life period, wearout period (= the period during which the codes become gradually obsolete, notwithstanding revisions, and must be completely abandoned and rewritten), series and parallel subsystems arrangement, etc. - Therefore, whenever a system of codes is subjected to changes, improvements, enlargements, etc., its reliability features must not be neglected and proper steps must be arranged to avoid its deterioration.

At present, the importance of code systems in economies is well appreciated and much effort is being attached to bringing the existing systems to perfection. This is an endless task, mainly because the potential code users are almost never happy with any change. On the other hand, the economic climate of actual history calls for unification in various fields of engineering activity. It is not necessary to explain why. Codes are now commonly considered to be a very powerful tool for achieving unification goals by successive steps. In Europe code harmonization efforts date back to the after-war period. In the construction industry they started somewhat later, in the early sixties. It must be noted that a similar course has been followed in the United States, where an interstate unification of codes is requested by economists and engineers, their motives being the same as those of their professional colleagues in Europe.

4.3 OPERATIONAL ASPECTS

It can be easily shown that RAP follows the general construction process; this consists of the following main sectors:

- planning which covers fundamental decisions on the purpose, location, size, and cost of a planned constructed facility; this section is usually managed by non-engineering bodies, such as owners, public authorities, etc.;

- design refers not only to the elaboration of design documents, but also to surveying, choice of the type of bearing structures, choice of other systems which are part of the constructed facility; here a consulting firm, architectural or structural, is the main performer;

- execution, supplied by contractors and subcontractors; it also includes transport of material and structural elements;

- quality control, inspection, and testing, where additional subcontractors, or also independent agencies, enter the process;

- maintenance and use, the responsibilities for which are carried by owners, or users, according to the type of facility and possible lease agreements.

The decision procedures which govern the mutual relationship between the foregoing sectors, and also internal arrangements within the sectors, can be called the operational factors of RAP. Thus, the aim of RAP in the operational field is to seek for optimum solutions for the procedures in each sector and to obtain good link-ups between the multitude of separate, simultaneous or successive, activities involved.

Operational defaults may have a very adverse effect on the reliability of any constructed facility; many examples of various types of failure due to such defaults can be given:

- insufficient geotechnical surveying can give biased information for the foundation design, with ensuing consequences;

- poor organization of design activities may cause overloading on designers who, consequently, are not able to analyse thoroughly the complex reliability features of the particular system;

- loss of control by the designer over execution of the design can lead to serious failure-producing mistakes;

- defaults in material supply can lead to a deterioration in material properties;

- inadequate maintenance can lead to early corrosion. Etc., etc.

It becomes evident that a spectrum of individuals and/or bodies must actively participate in RAP. In the majority of cases their contacts are very weak and, because of this, interface problems arrise. Transitions from one sector to the next sector are subjected to various types of decision.

For instance, in the selection of a consulting firm the following evaluation criteria are frequently used (see Gipe [5]);

- relevant experience,
- qualifications of firm and staff,
- special expertise,
- qualifications of sub-consultants,
- availability of key personnel,
- size of staff,
- current and projected workloads,
- geographic location.

Any decision taken in a certain sector of RAP can influence the reliability of the constructed facility <u>at any later moment of its service life</u>. A similar observation can be made concerning <u>non-decisions</u>, which must be expected whenever a certain aspect of the reliability is neglected, or even unknown. It often happens, for example, that no regular maintenance is prescribed by the design documents. This can result, after a couple of years, in poor behaviour of materials, deterioration, loss of durability, etc. Therefore, all sectors of RAP must be thoroughly linked-up, as closely as possible. The leading role of the designer is evident here, though his position is usually very ufavourable. Whereas the consulting firm is selected by means of the criteria mentioned above, the contractor is chosen on the lowest-bid principle, as a rule. Therefore, the designer does not usually know who will win the bidding in a particular case. This is a well known weak link in RAP, though it is economically fully justified, without any doubt, in any free market economy. However, whenever an outstanding facility (chemical process plant, powerplant, sporting stadium) is being planned, <u>design-construct firms</u>, covering two or more sectors of RAP and supplying full <u>construction management</u>, are preferred.

The decision making, at any point of RAP, is subjected to external influences which cannot be predicted (or sometimes even expected) by the decision-makers. This results in a specific <u>randomness of RAP</u>, the nature of which, unfortunately, has not yet been subjected to research. Though knowledge is minimum here, at present, it would be a gross mistake to approach RAP from only a deterministic position.

4.3.1 <u>Quality assurance and control process</u>

An important operational component of RAP is the <u>quality assurance and control process</u>, QAPC, which is usually presented in terms of two sub-components, viz. <u>quality assurance</u>, QA, and <u>quality control</u>, QC. The line between QC and QA is not clearly defined (see Kagan [6] where several interesting practical ideas on QC and QA are presented). The two sub-components overlap, more or less, or their contents can be mutually commuted. Similarly to RAP, the respective QACP has to start at the very beginning of the general construction process, and continues till the end of use of the constructed facility.

The difference between the two concepts, RAP and QACP, can be easily demonstrated on the reliability coefficients for material, γ_m . During code making, values of γ_m are established and based, theoretically or empirically, on a large set of assumptions. Now, the task of the QACP is to control and check the conditions, ensuring the validity of all those assumptions during the expected life of the facility. Obviously both the establishing of γ_m factors, and the control and checking of assumptions belong to RAP. Whereas γ_m factors are usually settled beyond the respective construction process, QACP runs through the whole life, T_0 , of the facility.

A constantly growing level of attention is now being paid to construction quality problems (see, e.g., Borges [7]). General concepts have not yet been stabilized, however. It happens that, under QACP, full reliability assurance is conceived, and the role of RAP as a decision process is not correctly understood, or recognized at all. Of course, much progress has been made as a result of computerization. There are many specialized software programmes available which treat the operational aspects of RAP, though subjective phenomena still affect most of the RAP sectors. The results of a computerized decision analysis must always be verified by independent means, based on engineering judgement, and possibly adjusted to the actual situation.

4.4 ECONOMIC ASPECTS

A general rule (perhaps platitudinous) can be stated: the more money is allocated to RAP, the higher the level of reliability of the particular constructed facility which is achieved. On the other hand, another rule rings true: the lower is the reliability level, the greater the possible costs involved with failures, repairs, redesign, litigation, etc. Obviously, a certain balance should be reached, in some way, so that the potential "failure cost" should equal "waste costs". Such a balance is the desired objective of any RAP, and, consequently, it may be ascertained that it is <u>mainly economic aspects which govern RAP</u>. However, the economic problems of RAP are far from being so straightforward as it first appears because not only monetary categories enter the considerations, but also some issues cannot be treated in economic terms at all. Therefore, whenever economic features of RAP are discussed a largely general meaning must be assigned to the concept of costs. It must be primarily understood that costs are equal to the benefits given up when a certain construction procedure is accepted. In this philosophy costs express:

- financial costs in terms of currency units,
- loss of life and limb,
- various general psychologistic values (emotional, moral, etc.),
- labour productivity.

It is hardly possible to unify all these cost branches in one scalar variable. Theoretically, it can be assumed that a certain unit of satisfaction, a "util" can be defined (see, e.g., Sievert [8]), on the base of which a <u>cost-benefit analysis</u> might be built-up, and the <u>opportunity cost</u> of a particular construction process covering RAP evaluated. This is, at present, obviously not posssible to do in general.

It should be noted that deviations from the two rules specified at the beginning of this paragraph can be observed whenever a simplified, and less expensive, design or construction procedure eliminates possible sources of human errors by reducing the complexity of particular operations (see, e.g., Stewart [9]). Unfortunately, this fact is not generally understood and very little benefits are drawn from it. - On the other hand, it must be kept in mind that simplifications to calculation models or inspection procedures, being intentionally set on the safe side as a rule, lead to oversizing of members, that is, to higher construction costs.

There are two principal areas where economic aspects control RAP. The first one refers to decisions on reliability levels, made in terms of variables affecting the system of proportioning and testing parameters; the other is represented by the economic climate of the particular country and/or period.

4.4.1 Decisions on reliability levels

In the reliability requirements (4.1) and (4.2) two basic types of quantities occur:

(a) facility variables which describe the physical and random properties of materials, dimensions, actions, environment, etc., on the basis of which the values of Z, P_f , β^{HL} , or others can be calculated, using an appropriate calculation model. It should be noted that P_f must be time-related: either a yearly, $\dot{P}_f$, or a summary, $\bar{P}_f$, value must be considered according to the type of solution used (the difference between $\dot{P}$ and $\bar{P}$ is only formal);

(b) reliability parameters which cannot be derived from the physical properties of the facility but are established by decisions - the reference period T_{ref} and the target failure probability, P_{ft} , that is, either $\dot{P}_{ft}$ or $\bar{P}_{ft}$, have to be determined by decisions based upon requirements of individuals or of social entities, upon economic analyses and, particularly, upon previous experience gained with similar facilities. Thus, both T_{ref} and P_{ft} are primary variables which can be viewed as having a similar initial importance in reliability analysis and design as, for instance, the Discount Rate and Money Supply have in the free market economies. By deciding on the appropriate reference period and the target failure probability, for a specified constructed facility, or for a whole class of constructed facilities, society (usually represented by a body of experts) takes on the responsibility for the number, and consequences, of possible failures of facilities designed for respective values of T_{ref} and P_{ft} .

While the nature of facility variables is physical and statistical, the nature of the two reliability parameters is obviously economic. Both the reference period and the target failure probability affect, on a large scale, through the ensuing values of the proportioning parameters, the consumption of material, labour, and energy.

4.4.2 Target life of constructed facility

In general solutions the reference period T_{ref} is usually assumed equal to 50 years. For refined solutions, however, this is not satisfactory at all and the target life of the respective facility, or another clearly specified period, should be taken as T_{wf} . Let us here concentrate on the problem of the target life in more detail.

The life of a constructed facility can be defined as the distance between two points in time: the moment of erection, $t = 0$, and the moment of demolition, $t = t_{dem}$ of the facility. This definition seems to be clear enough but it is not sufficient since the latter moment depends upon various factors and therefore must be specified in more detail.

The demolition of a facility may be caused, or provoked, by different circumstances. Basically, two types of demolition can be distinguished: foreseeable and unforeseeable. A foreseeable demolition is expected by both the designer and the owner and it is, in some implicit way, contained in the design, economic assessment, etc. On the other hand, unforeseeable demolitions are, as a rule, not considered at all, though it is commonly known that such demolitions may prevail over the foreseeable ones. Five principal reasons for demolition can be identified, and accordingly, five variants of the effective life, $T_{0,eff}$, can be defined, Tab. 4.1. Although in general, all factors governing the particular life variants can be considered random (even human decisions are subjected to randomness), only $T_{0,mt}$, $T_{0,ph}$, and $T_{0,bd}$ can be reasonably treated as random variables.

The effective life of a facility is obviously given by the minimum of the life variants shown in Tab. 4.1,

$$T_{0,eff} = \min(T_{0,mt}, T_{0,ph}, T_{0,ut}; T_{0,bd}, T_{0,sc})$$

In an effort to find the value of the target life, T_{0t} , this formula is not too helpful since each of the life variants is governed, as can be seen from Tab. 4.1, by substantially differing factors.

It is well known that the actual failure rate, λ , of buildings and structures, or in other words, the incidence of foreseeable demolitions is very small. Therefore, $T_{0,mt}$ does not appear in the designer's or owner's considerations (it is, however, not neglected in the areas of mechanical or electrical engineering). Neither do owners and designers, in their decisions, assume any unforeseeable demolitions, defining either $T_{0,bd}$ or $T_{0,sc}$; thus, only the physical life, $T_{0,ph}$, and the utility life, $T_{0,ut}$, remain as a basis for determining T_{0t} .

Table 4.1 - Lives of constructed facilities, CF

Type of demolition	Reason for demolition	Lives	Definition of T_0 or t_{dem}
Foreseeable	Random irreversible foreseeable failure of CF	Mathematical life, $T_{0,mt}$	$T_{0,mt}$ = reciprocal of the failure rate, λ
	Physical wear of CF	Physical life, $T_{0,ph}$	t_{dem} = point in time when maintenance or rehabilitation costs exceed an acceptable level
	Economic wear of CF (its further existence is not necessary)	Utility life, $T_{0,ut}$	t_{dem} depends on human decision
Unforeseeable	Random or non-random irreversible unforeseeable failure of CF caused by critical flaws or aberrations in the structure, loads or environment and resulting in break-down of CF	Break-down life, $T_{0,bd}$	t_{dem} is given by the break-down situation
	Social wear of CF	Social life, $T_{0,sc}$	t_{dem} is given by general economic situation, urban planning, political decisions, etc.

At the time of design, the value of $T_{0,ph}$ is unknown and it must be estimated from various properties of the constructed facility and the respective environment (material properties, actions properties, corrosive ambience, etc.). The value of $T_{0,ut}$ can often be specified more or less exactly but it is frequently exceeded, because of several reasons. Then, two points of view can be held in establishing the value of the target life:

- the owner should specify the required life of the facility, $T_{0,req}$, based mainly on $T_{0,ut}$;

- the <u>designer</u> should assume an <u>expected life</u>, $T_{0,exp}$, based either on $T_{0,ph}$ or on $T_{0,ut}$; if the designer relies on the owner's $T_{0,req}$, he usually takes $T_{0,exp} > T_{0,req}$ with a certain safety margin for the magnitude of which, however, no guidance exists at present.

During the use of a constructed facility the circumstances considered by the owner or designer may change, and so it finally becomes

$$T_{0,eff} \leq \text{or} > T_{0,req}, \quad T_{0,eff} \leq \text{or} > T_{0,exp}$$

Of course, the influence of random phenomena may result in

$$T_{0,mt} < T_{0,req}, \quad T_{0,mt} < T_{0,exp}$$

<u>Table 4.2</u> - The values of life (years) obtained from the opinions of civil engineers

Constructed facility	Material	Life $T_{0,ph}$	Life $T_{0,sc}$
Residential buildings	Masonry	112	60
	Concrete	115	66
Single-storey industrial buildings	Concrete	88	46
	Steel	61	33
Highway bridges	Concrete	108	54
	Steel	81	39
Gravity dams	Concrete	258	259
	Earth	217	202
Grain silos	Concrete	102	78
	Steel	52	73
Tanks	Concrete	84	68
Chimney stacks	Masonry	85	80
	Concrete	92	66
	Steel	31	47
Cooling towers	Concrete	93	43
	Steel	42	22
Weekend chalets	-	55	32

When establishing values of the target life, T_{0t}, expert opinion, as well as economic considerations, must be used; such an approach has been used by the Author. During an investigation of the problem, 46 outstanding civil engineers from different circles of construction in Czechoslovakia gave their estimates of $T_{0,ph}$ and $T_{0,sc}$ for various types of constructed facilities. Some results of the inquiry are shown in Tab. 4.2. No economy experts were involved at this stage, and so it can be said that the values of $T_{0,ph}$ are, in fact, close to the designer's $T_{0,exp}$. For simplicity, the table shows sample means only. The sample range of opinions, however, was surprisingly narrow for most types of facilities.

In the next step of the solution, economic criteria were taken into account, considering the depreciation period T_{dep} specified for buildings and structures in a Czechoslovak legal document. Assuming that the owner's $T_{0,req}$ should be by about 20 to 30% greater than the respective depreciation period and using also the expert inquiry results, values of target life, T_{0t}, were finally established, Tab. 4.3.

<u>Table 4.3</u> - Guidance values of the target life, T_{0t}, specified in Czechoslovak code CSN 73 0031-88 (years)

Constructed facilities	T_{0t}
<u>Buildings</u>	
housing	100
industry	60
mining	50
energy supply	30
agriculture	50
hydrotechnics	80
temporary	15
<u>Structures</u>	
towers	40
tanks, bunkers	80
bridges	100
highways, general structure	100
rigid surface	25
non-rigid surface	15
railroads, general	120
bed	40
dams	120
tunnels, underground facilities	120

4.4.3 Target failure probability

Much attention has been paid to the values of P_{ft} but results have been rather poor until now. Tables of suggested values of P_{ft} are shown, in various general codification documents, but

a common consensus on P_{ft} has not yet been reached. Evidently, the problem of the target failure probability is more intricate than it appears, and it is definitely more complicated than that of the target life, discussed in the foregoing paragraphs. Whereas T_{0t} is a meaningful, independent quantity, which, in a way, is testable and can be verified by experience, P_{ft} is a value that is difficult to conceive and to be checked by common designers, contractors, or many other participants of RAP. Further, whenever a value of P_{ft} is given, the calculation model, or rather a complex system of calculation models, supplying the failure probability P_f must also be simultaneously defined. It has been graphically shown (see Grimmelt and Schueller [10]) that even for very simple, textbook structures, with clearly defined properties, subjected to clearly defined actions, a wide spectrum of P_f values can be obtained by using different calculation methods. The reason for the observed discrepancies is obvious: any calculation model proposed at present is only a very rough approximation of actual system behaviour. - On the other hand, since no connection between T_{0t} and the calculation model exists, the T_{0t} cannot be model-dependent.

Among code-makers there exists a natural psychological reluctance to give definite values of P_{ft} and to accept the idea that a certain proportion of constructed facilities will fail. This reluctance to fix P_{ft} strengthens with the growing potential damage consequence of a failure.

For these reasons, any recommended values of P_{ft} must be viewed with utmost caution and always in the context of the set of factors affecting the reliability. This fact, however, should not prevent us from discussing briefly several ways of arriving at P_{ft} values.

It is recalled that two levels of target failure probabilities need to be specified: one for serviceability failures, and the other for ultimate failures. The P_{ft} values for these levels may differ by many orders of magnitude. This is obviously due to the well known attitudes of the public, regarding the two types of failure. As a guidance it can be said that for serviceability limit states the summary value $\bar{P}_{fts}$, referred to, say, T_{0t} = 70 years, is in the range of 10.E-1 to 10.E-3, whereas for ultimate limit states it should be considerably less, say $\bar{P}_{ftu} \in$ (10.E-5 to 10.E-8). However, it must be kept in mind that very small probabilities, that is, 10.E-6 and less are very, very doubtful, untestable numbers (cf. on this issue, but in a substantially different environment, Feynman [11] on investigating the space shuttle Challenger disaster).

Four fundamental methods of fixing P_{ft} can be distinguished:

(a) Recalculation method. This method consists simply in the analysis of an existing system by means of a specified probabi-

where A, B, C, D and E are the parameters to be determined by a nonlinear least square procedure based on actual earthquake records. The main advantage of Eq. 5.18 is that, by adjusting D and the ratio of B to E, the rise and fall of most recorded energy functions can be reproduced closely.

To characterize the frequency content change with time, the zero crossing rate of the recorded accelerogram is used. Assuming that the mean value of the total zero crossings of the recorded accelerogram up to time t, denoted by $\mu_0(t)$, is a continuous differentiable and non-decreasing function of time, a functional form for $\mu_0(t)$ is postulated and its parameters are estimated by a least square procedure (Yeh and Wen, 1989). The functional form chosen for $\mu_0(t)$ and $\phi(t)$ are

$$\mu_0(t) = r_1 t + r_2 t^2 + r_3 t^3 \tag{5.19}$$

$$\phi(t) = \frac{\mu_0(t)}{\mu_0'(t_0)} \tag{5.20}$$

where r_1, r_2 and r_3 are the parameters to be determined, t_0 corresponds to the starting time of significant excitation, and the prime denotes the time derivative.

To estimate the remaining parameters in the Kanai-Tajimi or Clough-Penzien filters, i.e., Eqs. 5.5 and 5.6, the recorded accelerogram is first scaled in intensity and time by the already identified modulation functions I(t) and $\phi(t)$ such that it is approximately stationary. The power spectrum of the reduced accelerogram is then calculated by the standard method and the parameters ω_g, ζ_g, ω_f and ζ_f are identified by a system identification procedure (Yeh and Wen, 1989).

In case not only the dominant frequencies but also the forms of power spectra of ground motion change significantly in different time intervals, the original accelerogram is decomposed into several components. Each of them has banded frequencies and is modeled separately. The original accelerogram is then obtained by a linear combination of the simulated components. Details are available in Yeh and Wen (1989).

Random Vibration Analysis of Inelastic Structures

Structures under strong motion earthquake excitations often go into inelastic range, and the restoring forces become nonlinear and hysteretic. The structure may deteriorate in stiffness and/or strength as it undergoes repeated, large amplitude oscillations. This loss of stiffness (hence, lengthening of structural period) may further coincide with the shift in frequency content of ground excitation and magnify the response. Effects of time-varying frequency content of ground excitation on the structural response may be studied analytically using the new

analyses are extremely difficult and until now the respective methods have not escaped bounds of textbooks. The main difficulty is not in the optimization procedures, which are now attainable with present computer hardware and software, but in the calculation models.

4.4.4 Economic climate

It can be ascertained that in RAP considerations the owner-designer-contractor chain appears, affecting the economy of the particular construction process in detail, and consequently affecting RAP. Any savings achieved by one of the members of this chain may typically bring loss to himself or herself, and to other members also. Thus, RAP is subjected at this level, to short-run economic considerations dominated by the current economic situation of the respective environment.

In the particular sectors of RAP some rudimentary cost-benefit analyses can be made by the designer. - He/she should consider, e.g., the opportunity cost of changing a certain value of the target failure probability, P_{ft1} , to another value, P_{ft2} . Or, the contractor can choose some patent procedures which are supposed to decrease the outlays of the bid and, analogously, the owner may abandon some maintenance action, accepting the risk of an earlier decay of the facility. Many examples of good and bad decisions can be given here.

Now, the economic problem of RAP extends beyond the owner-designer-contractor chain because the dynamics of the national, or even world economy may exert considerable influence upon RAP and the resulting reliability level, in many ways. In the first place, any economic restrictions imposed on the construction industry by recession, by political decisions, or by other investment-unfriendly factors are accompanied by tendencies to keep down bid budgets, that is, to diminish costs. As a rule, QACP is affected initially, then material quality, maintenance procedures, etc. Further discussion of these phenomena would be superfluous here.

Conversely, an economic expansion may produce very similar effects upon the reliability of constructed facilities, to those brought up by recession. Since the availability of resources, that is, of labour, enterpreneurship, capital, and natural resources becomes limited at expansion periods, the construction industry cannot follow the development, and, as a rule, the shortage of resources is supposed to be counterbalanced by pressures on labour availability. Under such circumstances, not too much attention is being paid to the education of labourers as a mean of improving the resource situation. The subsequent negative impact on reliability is obvious again.

Similar phenomena affecting reliability during recession or expansion may also be observed during any inflation period, which may string along with both the recession or the expansion. Time factors apply here, since construction projects are to be completed in the shortest time and for the least initial cost (see Carper in Forensic Engineering [16]), which tends to escalate failure rates.

The economic factors of RAP are essentially the same in all market economy systems, as well as in command systems. Whenever basic economic laws are distorted by non-economic aberrations, the reliability of constructed facilities suffers. Many examples of spectacular failures, with an economic background, can be given from various countries.

4.5 LEGAL ASPECTS

The principles of structural mechanics are identical worldwide; the rules of execution and workmanship deviate slightly from country to country. Economic rules differ according to historical and political situations in the particular region but in framework of various economic systems they still follow analogous principles. In contrast, the laws Man has imposed upon himself show a very diversified pattern. Owing to this diversity, legal aspects of reliability assurance cannot be uniform. Nevertheless, <u>two basic effects of the legal climate</u> on the reliability of constructed facilities may be identified in any country, and in any socio-economic system:

- regulatory effects,
- deterrent effects.

In the first case more or less complicated assemblies of laws, rules, codes, etc., formulate <u>mutual relations between the participants of the construction process</u>, and their <u>relations to the society</u>, usually represented by authorities on various levels (national, provincial, cantonal, municipal, local, or others). Further, other regulations specify <u>performance characteristics</u> of construction products, buildings, bridges, or other facilities. Design and execution codes, which have already been mentioned here, also belong to this family though they are primarily conceived as tools of mutual understanding and unification. When all the particular regulations are carefully followed, everything runs smoothly, no legal problems arise and reliability is well under control.

The spectrum of the regulations which are to be respected is very wide. It contains, first, laws which have at first glance nothing in common with engineering reliability - criminal, civil, labour, tort law, maybe also others, and second, many regulations which refer directly to constructed facilities and which treat fire protection, occupational safety, and other aspects of good performance.

Most of these documents contain <u>sanction clauses</u> saying what punishment is presumed when the regulatory clauses are not respected. A direct deterrent effect of regulations thus enters RAP. However, the main deterrent effects of a legal character are direct.

The threat of <u>litigation on construction issues</u> has expanded during recent decades. The reasons for this tendency are perhaps known to lawyers and politicians but they often are not fully clear to engineers. What, however, <u>is</u> known is that fear from being involved in a court trial has an extremely favourable influence upon the quality assurance and control with the designer, and with the contractor as well. Both these parties are usually well aware of problems following a structural failure which had been subjected to attention of public during the investigative period and, then, during litigation. The subsequent loss of credibility is often more expensive than the damage paid by the respective liable party. Moreover, firms which are not found responsible also suffer owing to the fact that the public becomes suspicious to all who are involved, in some way, in a failure case. Whenever possible, the decisions of a private judge or arbitrator are preferred to court trials.

It can be observed, in the long run, that QACP costs have increased to about 1 to 3% of the project cost, on the average. This is a definitely reasonable investment since in countries where the importance of QACP is not understood and appreciated, various reworking costs can reach a 5% level, or even higher. The merits of a systematic QACP do not refer to catastrophic failures only. It is equally important to prevent, or remove, minor mistakes which can eventually bring major trouble to the owner or which can simply delay the execution, so that the time schedule is not complied with. Delays are always costly, and contractors happen to be sued by owners for money lost because of a late start of operations of the particular facility. Here, insurance companies may participate in the construction process, but they always check whether the reliability is sufficiently covered.

It can thus be maintained that the juridical aspects of the construction process, and also of the post-construction period, have a definitely positive bearing on the level of reliability of constructed facilities. There is an exception to every rule, however. It happens that new design procedures, and new technologies, are being stubbornly resisted by the profession, for a long period, owing to insufficient experience with the new techniques. Managers are afraid of getting sued for potential losses caused by some burn-in errors committed by the personnel when using the new solution (cf. the problems arising in the United States with limit state design code for steel structures, as descibed by Burns and Rosenbaum [4]).

4.6 CONCLUSIONS

The objectives of the foregoing paragraphs were to show the complexity of reliability assurance for constructed facilities, and the main features of the whole process. The various problems could not be described in great depth, owing to the lack of space, nor could every particularity be discussed or suggested. Therefore, many issues are left to the Reader for his study and consideration, such as, for example, that of forensic engineering which is a discipline closely related to RAP (see Forensic Engineering [16]). In fact, what was presented in the preceding pages is a brief summary of some theoretical and empirical features of the reliability assurance. It is difficult to draw conclusions from a summary, but nevertheless it may be noted that the complexity of RAP does not require only a thorough training in sophisticated probability-based calculation methods but also a good understanding of operational procedures, economics, and law. Of course, the assistance of experts on individual practical issues is always necessary but the overall management of RAP must be performed by people who have a good feeling for the complex. It can be said that a new profession, that of the <u>Reliability Engineer</u>, is just emerging at present, and that, in the future, reliability engineers will act as advisers to designers, contractors, and public authorities in most major construction projects, as well as in the everyday construction problems of the creative community.

REFERENCES

1. Taoka, G.T.: Civil engineering design professors should be registered engineers, Journal of Professional Issues in Engineering, 115 (1989), No.3, 235-240.

2. Tichy, M.: The Nature of Reliability Requirements, in: Methods of Stochastic Structural Mechanics, Proc. of the 2nd International Workshop on Stochastic Methods in Structural Mechanics, Universita di Pavia (Ed. F. Casciati and L. Faravelli), SEAG, Pavia 1985, 153-165.

3. Madsen, H.O., Krenk, S., and Lind, N.C.: Methods of Structural Safety, Prentice-Hall, Englewood Cliffs, N.J., USA 1986.

4. Burns, G. and Rosenbaum, D.B.: Steel design's reluctant revolution, Engineering News Record, November 9, 1989, 54-60.

5. Gipe, A.B.: How to select a consulting engineer, Engineering News Record, 222 (1989), No. 24, CE-10 to CE-12.

6. Kagan, H.A.: Practical quality-controlled construction, Journal of Performance of Constructed Facilities, 8 (1989), No. 3, 191-198.

7. Borges, J. Ferry: Qualidade na conctrucao, Curso 167, Laboratorio Nacional de Engenharia Civil, Lisbon 1988.

8. Sievert, D.M.: Economics - Dealing with Scarcity, Glengarry Publishing, Waukesha, Wisconsin, USA 1989.

9. Stewart, M.G.: Safe load tables and the human dimension, Steel Construction (Australia), 24 (1990), No. 1, 2-12.

10. Grimmelt, M.J. and Schueller, G.I.: Benchmark study on methods to determine collapse failure probabilities of redundant structures, Structural Safety, 1 (1982), No.2, 93-106.

11. Feynman, R.P.: "What Do You Care What Other People Think?", Bantam Books, New York 1985.

12. Kuhlmann, A.: Introduction to Safety Sciences, Springer Verlag, New York 1985.

13. Bennett, R.: Loss of life expectancy and other measures of risk, Risk Abstracts, 6 (1989), No. 1, 1-5.

14. Diaz Padilla, J. and Robles, F.: Human Response to Cracking in Concrete Slabs, in: Publication SP30, American Concrete Institute, Detroit 1971.

15. Needleman, L.: Methods of Valuing Life, in: Technological Risk, Proc. of a Symp. on Risk in Technologies (Ed. N.C. Lind), University of Waterloo Press, Waterloo, Ontario, Canada 1982, 89-99.

16. Forensic Engineering (Ed. K.L. Carper), Elsevier, New York 1989.

Chapter 5

APPLICATION OF NONLINEAR STOCHASTIC DYNAMICS AND DAMAGE ACCUMULATION IN SEISMIC ENGINEERING

Y. K. Wen
University of Illinois, Urbana, IL, USA

5.0 INTRODUCTION

The response of a structure, in particular in the inelastic range, is highly dependent on the ground excitation intensity, duration as well as frequency content and its evolution with time. Strong ground motions need to be modeled as nonstationary random processes. Although methods have been proposed to generate nonstationary random processes with evolutionary power spectra, the required parameters are difficult to estimated and the methods are difficult to implement in random vibration response studies. This chapter introduces a new nonstationary ground motion model and its applications. It can be efficiently used in Monte Carlo simulations as well as random vibration response studies. Methods for parameter estimation are also given.

Methods for random vibration analysis of nonlinear systems are then introduced with special emphasis on the method of equivalent linearization and time domain approach using a differential equation model for the hysteretic systems. Ample references are given for more details of this versatile method. Application of the random vibration method to response and damage analysis of actual reinforced concrete and masonry structures under seismic loading is also given which demonstrates the accuracy and usefulness of this method for practical problems. Finally, the method of fast integration for reliability analysis of time variant systems with uncertain system parameters is given, which allows one to evaluate the overall system reliability efficiently.

5.1 MODEL OF NONSTATIONARY EARTHQUAKE GROUND MOTION AND APPLICATIONS

It is well understood from study of recorded accelerograms that earthquake ground motion is generally a nonstationary random process in which both intensity and frequency content change with time. For excitations which can be modeled by stationary random process, the auto-correlation and spectral density functions provide useful information to assess system response and reliability. However, for a nonstationary random process, the concepts of auto-correlation and spectral density functions become more involved; for example, the auto-correlation function depends on two specific time instants and there is no unique definition of the associated time-varying power spectrum. Evolutionary, physical and instantaneous power spectra are all legitimate descriptions of time-variant frequency content and are suitable for certain applications. An excellent review of these time-variant power spectral density functions is given in Mark (1986).

Recent models are mostly based on an evolutionary power spectrum, which gives an accurate description of both frequency and intensity changes with time. Estimation of parameters in these models or implementation in analytical random vibration analyses, however, are not easy matters. On the other hand, a commonly used piece-wise stationary model which consists of distinct stationary random processes in consecutive time intervals is difficult to justify physically.

In this chapter, a new stochastic model of ground excitation is proposed. It is a modification of the amplitude and frequency modulation random processes proposed by Grigoriu, et al. (1988). Methods for estimating the model parameters based on actual accelerograms are also given. The new model is suited for simulations and time history analyses as well as analytical random vibration analyses. The frequency content of the simulated ground excitation changes with time continuously and smoothly. It is compatible with a method for the random vibration response analysis of inelastic structure developed by Wen (1989) and Yeh and Wen (1990).

Frequency Modulation

Consider a stationary random process $Y(\phi)$ which has the following spectral representation

$$Y(\phi) = \int_{-\infty}^{\infty} e^{i\omega\phi} dZ(\omega) \tag{5.1}$$

where $Z(\omega)$ is a random process in ω and has orthogonal increments, i.e., $E[dZ(\omega_1)dZ(\omega_2)] = 0$, for all $\omega_1 \neq \omega_2$. Physically, the spectral density function of a stationary random process is defined by averaging the power at each frequency over all the realizations, i.e., $S_{YY}(\omega)d\omega = E[|dz(\omega)|^2]$. The auto-correlation function of $Y(\phi)$ based on Eq. 5.1 is

$$R_{YY}(\tau) = \int_{-\infty}^{\infty}\int_{-\infty}^{\infty} e^{i\omega_2\phi_2 - i\omega_1\phi_1} E[dZ(\omega_1)dZ(\omega_2)]$$

$$= \int_{-\infty}^{\infty} e^{i\omega\tau} S_{YY}(\omega)d\omega \tag{5.2}$$

which is the Fourier transform of the spectral density function. Equation 5.2 is the well known Wiener-Khintchine relationship for a stationary random process.

Now let $X(t)=Y(\phi(t))$ and $\phi(t)$ be a smooth, strictly increasing function of t. For the simplest case, where $\phi(t)$ is a linear function of t, the process X(t) can be shown to be stationary but with a different time scale. If $\phi(t)$ is a nonlinear function of t, X(t) is no longer stationary because of the continuously changing time scale effect. The auto-correlation function of X(t) at a given time instant t can be shown to be

$$R_{XX}(t,\tau) = E[X(t+\tfrac{\tau}{2})X(t-\tfrac{\tau}{2})]$$

$$= \int_{-\infty}^{\infty}\int_{-\infty}^{\infty} e^{-i\omega_1\phi(t+\frac{\tau}{2}) - \phi_2(t-\frac{\tau}{2})]} E[dZ(\omega_1)dZ(\omega_2)]$$

$$= \int_{-\infty}^{\infty} e^{i\omega[\phi(t+\frac{\tau}{2}) - \phi(t-\frac{\tau}{2})]} S_{YY}(\omega)d\omega$$

$$\approx \int_{-\infty}^{\infty} e^{i\omega\phi'(t)\tau} S_{YY}(\omega)d\omega$$

$$= \int_{-\infty} e^{i\bar{\omega}t} \frac{1}{\phi'(t)} S_{YY}(\frac{\bar{\omega}}{\phi'(t)})d\bar{\omega} \tag{5.3}$$

where $\bar{\omega}=\phi'(t)\omega$ and $\phi'(t)=d\phi/dt$. It is noted that in Eq. 5.3 the difference in ϕ is replaced by the time lag τ multiplied by the first derivative of $\phi(t)$. This is justified for a smooth function $\phi(t)$ and is verified for most earthquake ground excitations (Yeh and Wen, 1989). From the definition of an instantaneous power spectrum, one can see that Eq. 5.3 is the corresponding Wiener-Khintchine relationship of a nonstationary random process, and the instantaneous power spectral density function of X(t) is

$$S_{XX}(t,\omega) = \frac{1}{\phi'(t)} S_{YY}(\frac{\omega}{\phi'(t)}) \tag{5.4}$$

In summary, Eq. 5.4 relates the power spectra of the two processes $X(t)$ and $Y(\phi)$, and the first derivative of $\phi(t)$ plays an important role in this relationship. Note that such frequency modulated random processes do not necessarily have evolutionary power spectra (Grigoriu, 1988).

Linear Filtering with Intensity and Frequency Modulation

To model the frequency content of earthquake ground motion, a Gaussian filtered white noise with a Kanai-Tajimi or Clough-Penzien spectrum are frequently used. The time-varying intensity and frequency content can be incorporated into a random process with the Kanai-Tajimi or Clough-Penzien types of spectrum, by using a time-variant linear filters as follows.

Consider the random process generated by passing a Gaussian white noise $\zeta(t)$ of constant spectral density S_o through a filter given by

$$\ddot{x}_g + 2\zeta_g \omega_g \dot{x}_g + \omega_g^2 x_g = -\zeta(t) \tag{5.5}$$

where ω_g and ζ_g are constant parameters. The composite response $\zeta_{KT}(t)=2\zeta_g\omega_g\dot{x}_g+\omega_g^2 x_g$ is stationary and has the well known Kanai-Tajimi spectrum, which will be denoted by $S_{KT}(\omega)$. To remove the low frequency content, one can pass the composite response through a high pass filter

$$\ddot{x}_f + 2\zeta_f \omega_f \dot{x}_f + \omega_f^2 x_f = -\zeta_{KT}(t) \tag{5.6}$$

and the resultant acceleration $\ddot{x}_f$ is also stationary and has a power spectral density which goes to zero according to ω^4 as $\omega \to 0$ and is referred to as the Clough-Penzien spectrum $S_{CP}(\omega)$. Note that if the input stationary white noise in Eq. 5.5 or the output responses are multiplied by a deterministic intensity envelope $I(t)$, the resulting processes are known as uniformly (or amplitude) modulated random processes, which have been used extensively as nonstationary random excitations.

Assuming now the independent variable to be ϕ and introducing a time scaling function $\phi(t)$, by chain rule, Eqs. 5.5 and 5.6 reduce to

$$\frac{d^2 x_g(t)}{dt^2} + \left[-\frac{\phi''(t)}{\phi'(t)} + 2\zeta_g \omega_g \phi'(t)\right]\frac{dx_g(t)}{dt} + [\omega_g \phi'(t)]^2 x_g(t)$$

$$= -[\phi'(t)]\, \zeta(\phi(t)) \tag{5.7}$$

$$\frac{d^2x_f(t)}{dt^2} + \left[-\frac{\phi''(t)}{\phi'(t)} + 2\zeta_f\omega_f\phi'(t)\right]\frac{dx_f(t)}{dt} + [\omega_f\phi'(t)]^2x_f(t)$$

$$= -2\zeta_g\omega_g\phi'(t)\frac{dx_g(t)}{dt} - [\omega_g\phi'(t)]^2x_g(t) \tag{5.8}$$

It should be noted that because of the transformation of variables the power spectral density of $\zeta(\phi(t))$ in Eq. 5.7 at a given time t is $S_0/\phi'(t)$. Applying the same transformation of variables, the responses $\zeta_{KT}(t)$ and $\zeta_{CP}(t)$ are expressed in terms of t. The spectral density function of the composite response $2\zeta_g\omega_g\dot{x}_g/\phi'(t)+\omega_g^2x_g$ at a given time t is

$$S(t,\omega) = \frac{1}{\phi'(t)} S_{KT}\left(\frac{\omega}{\phi'(t)}\right) \tag{5.9}$$

and that of the response $2\zeta_f\omega_f\dot{x}_f/\phi'(t) + \omega_f^2x_f + 2\zeta_g\omega_g\dot{x}_g/\phi'(t) + \omega_g^2x_g$ is

$$S(t,\omega) = \frac{1}{\phi'(t)} S_{CP}\left(\frac{\omega}{\phi'(t)}\right) \tag{5.10}$$

Integrations of Eqs. 5.9 and 5.10 with respect to ω indicate that the variance of a frequency modulated random process does not change with time. Thus, if the input white noise in Eq. 5.7 is multiplied by an intensity envelope I(t), the variance of the resulting amplitude and frequency modulated random process is controlled by I(t) only.

Ground Motion Model

In summary, earthquake ground excitation $\xi(t)$ is expressed as

$$\xi(t) = I(t)Y(\phi(t)) \tag{5.11}$$

where I(t) is a deterministic intensity envelope controlling the amplitude of the resultant process, and $Y(\phi(t))$ is the frequency modulated random process as given by Eqs. 5.7 and 5.8 with a zero mean and a unit variance. The parameters required for the model include those of the filters and the two time-varying modulation functions I(t) and $\phi(t)$. It is pointed out that the ground excitation model given by Eqs. 5.7, 5.8 and 5.11 can be easily generated on computer. Perhaps more importantly, it is compatible with the recently developed and proven time-domain random vibration method for inelastic systems. A summary of this method can be found in Wen (1989).

In the more general case, the earthquake ground motion at a given point may be modeled as a multi-directional random process. In this study, the vertical and rotational motion are not considered. The two horizontal accelerations along the excitation principal axes can be written as

$$\xi_1(t) = I_1(t)Y_1(\phi(t)) \tag{5.12}$$

$$\xi_2(t) = I_2(t)Y_2(\phi(t)) \tag{5.13}$$

where $I_1(t)$, $I_2(t)$ are deterministic intensity envelopes, and $Y_1(\phi(t))$, $Y_2(\phi(t))$ are uncorrelated, frequency modulated random processes given by Eqs. 5.7 or 5.8 with zero means and unit variances. In general, the principal axes of the structure do not coincide with those of the ground excitation. Let the orientation angle of the structure with respect to the principal axes of the excitation be θ. The corresponding accelerations along the structural principal axes are given as

$$\xi_x(t) = \xi_1(t)\cos\theta - \xi_2(t)\sin\theta \tag{5.14}$$

$$\xi_y(t) = \xi_1(t)\sin\theta + \xi_2(t)\cos\theta \tag{5.15}$$

Parameter Estimation

To estimate the required parameters and time-varying modulation functions in the proposed ground motion model, the following procedure is proposed.

The expected energy function of ground excitation is given by

$$E[W(t)] = \int_0^t E[\xi^2(\tau)]d\tau \tag{5.16}$$

which, according to the proposed model, is

$$E[W(t)] = \int_0^t I^2(\tau)E[Y^2(\phi(\tau))]d\tau$$

$$= \int_0^t I^2(\tau)d\tau \tag{5.17}$$

where $I(t)$ and $Y(\phi(t))$ are defined in Eq. 5.11. The required parameters in the intensity envelope $I(t)$ are chosen so that the expected energy function based on the model fits that of a recorded accelerogram in a least square sense. After repeated trials, the following functional form for the intensity envelope proves to be most flexible and accurate,

$$I^2(t) = A \frac{t^B}{D+t^E} e^{-Ct} \tag{5.18}$$

lity-based method. It is assumed that the value of the failure probability, P_{ft}, resulting from this solution is equal to P_{ft} for that particular system. If a large set of systems is subjected to such analysis, a set of P_{ft} values is obtained and, after some considerations, the most acceptable value of P_{ft} is applied in the design of future facilities, or in the derivation of proportioning parameters for codes. Clearly, the recalculation method is based on past experience with constructed facilities which already are in current use, and, in this way, P_{ft} depends upon these facilities themselves. The method can be applied for both the ultimate and the serviceability limit states. Unfortunately, it is model-dependent.

(b) Analogy method is based on the evaluation of other phenomena of a catastrophic nature. Therefore, ultimate failure probability, P_{fu}, is studied by this method. It is suggested that the target value for ultimate limit states, P_{fut}, should be derived, for example, from the probability that an individual would be accidentally killed on his/her way home from his/her office. Many other possibilities have been offered (see, e.g. Kuhlmann [12], Bennett [13]). Several drawbacks of the analogy method can be shown, but, on the other hand, the method gives open space for engineering judgement.

(d) Discomfort method. Any defect or fault in a constructed facility creates uneasy feelings amongst users and owners. An appropriate analysis of the attitudes of individuals or groups in assumed or real failure situations can lead to reasonably well-founded, model-independent target failure probabilities. Methods of attitude evaluation are elaborated by applied psychology but their use in RAP is rather uncommon (for a unique example see Diaz and Robles [14]). The discomfort method becomes particularly suitable when the definition of the respective limit state is fuzzy (deflection limit state, crack-width limit state, and others). It is not of too much help, however, in the domain of ultimate limit states.

(d) Optimization methods are the most exact of those methods aiming at the target failure probabilities: they usually deal with an economic analysis of the constructed facility system, in time. The costs of design, execution, quality control and maintenance of a constructed facility, and also damage caused by possible failures of the facility are combined into an appropriate objective function where separate P_{ft} values for possible types of failure must appear as variables. Then, as objectives of the solution, the values of P_{ft} probabilities can be found by minimizing the total costs, under defined constraints. The objective function can also be written in terms of energy, or material consumption, or in terms of the losses due to failures only, but the optimization principle remains economic (see Needleman [15]). The merits of the method are rather eclipsed by the simple fact that human lives, in civilized societies, and under normal conditions, are not subjectable to optimization, and whenever ultimate failures which may involve loss of lives are captured by the objective function, the whole solution becomes doubtful. This problem is dealt with by proposals for minimizing the mortality rate due to structural failures and determining the P_{ft} under such approach. Yet, all optimization

ground motion model. By method of equivalent linearization, response statistics including the r.m.s. displacement and velocity, ductility and energy dissipation of unsymmetric multi-story buildings, with or without deterioration, can be obtained. Details can be found in Yeh and Wen (1989).

The validity of the proposed ground motion modeled is examined by comparison of simulated with actual ground accelerations as well as inelastic structural responses. Also examined is the effect of nonstationarity of the ground motion on structural response.

Figure 5.1 compares the recorded with the simulated accelerograms based on the proposed method for the Venture accelerograms, 1971, where there is obvious changes in frequency content with time. Figure 5.2 shows the ductilities of SDOF inelastic systems under simulated earthquake excitation (solid square, sample size 10) and comparisons with those calculated from the actual accelerogram (hollow square). The mean values are shown by the solid lines. The scatter shows the variability due to the inherent randomness of the excitation since all the excitations are statistically equivalent. It is clear that design decisions should not be based on the result from a single accelerogram.

Figure 5.3 shows the effects of time-variant frequency content of ground excitation on responses of structures with different pre-yielding natural frequencies. The rate of change of dominant frequency of ground excitation comes from the estimation of Venture accelerogram, 1971, which shows significant lengthening of the ground excitation period in the later stage of the excitation. The thin and thick lines in the figure correspond to the analytical results of inelastic structures under ground excitation with time-invariant and time-variant spectral contents, respectively. Monte Carlo simulation results are also shown for comparison. It is seen that, for soft structures (with fundamental natural frequency less than 1 cps), expected energy dissipations and ductilities are all significantly magnified by changing frequency content of ground excitation. The magnification factor could be as high as 2 or 3. However, for very stiff structures, the effect of time-variant frequency content of ground excitation is small.

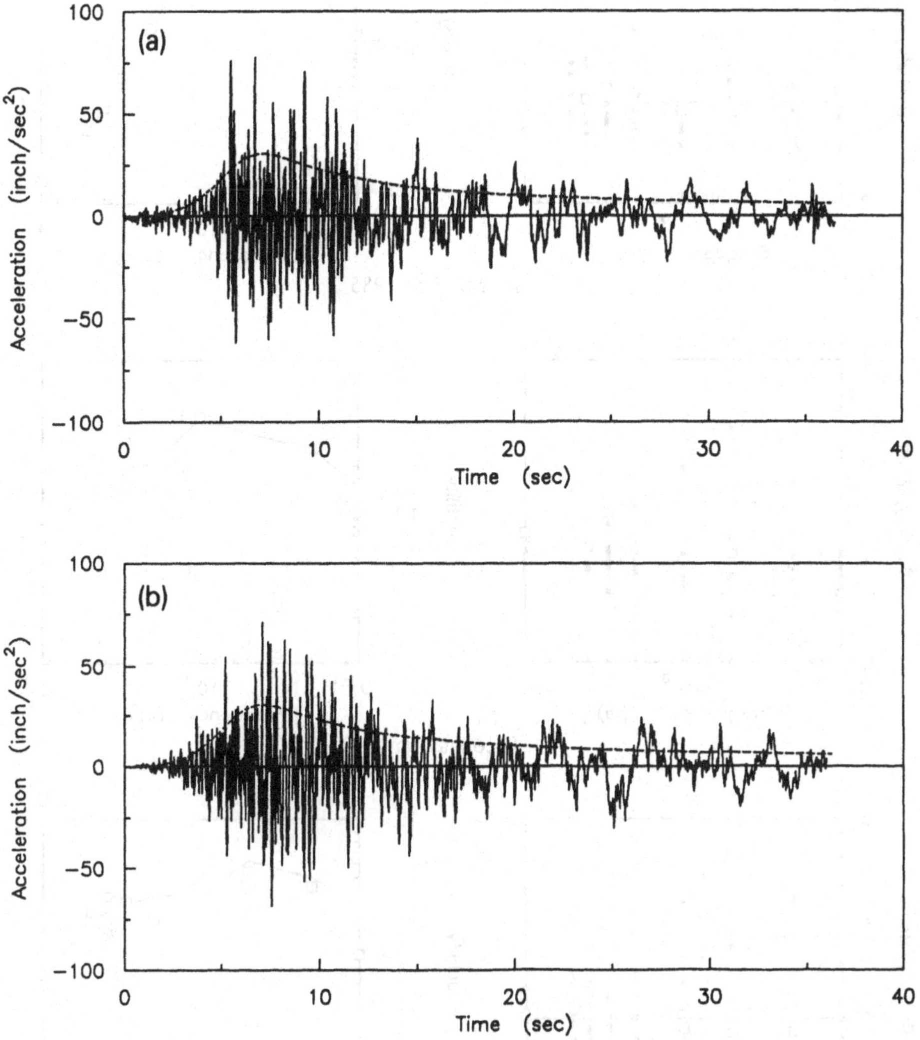

Fig. 5.1 Comparison of recorded (top) and simuilation (bottom) accelerogram (Ventura, 1971)

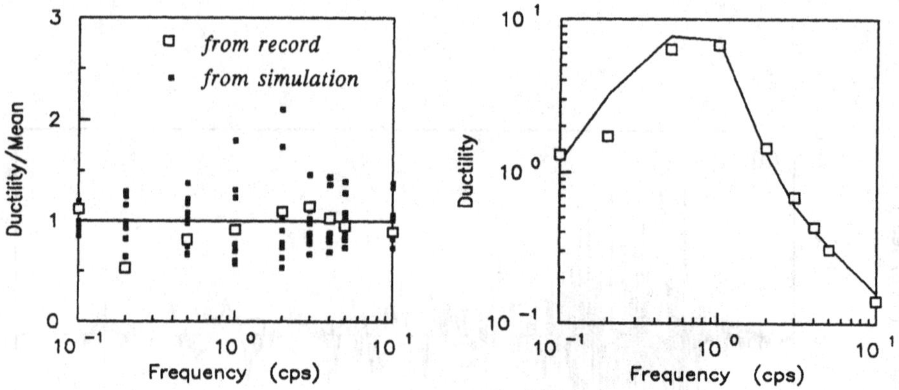

(a) Mexico, 1985

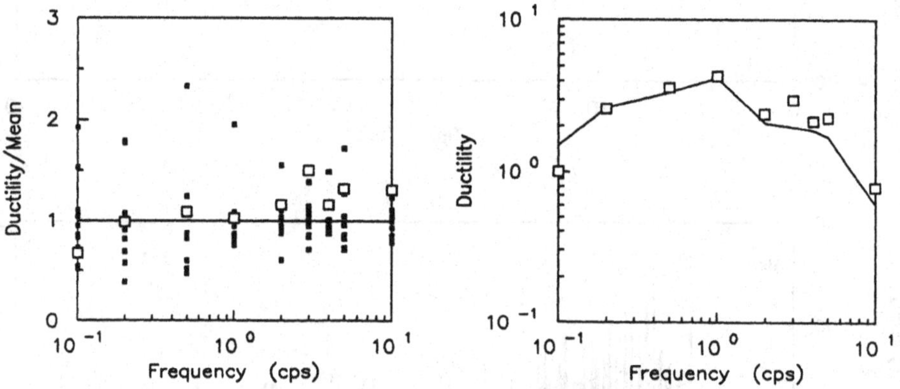

(b) Pacoima Dam, 1971

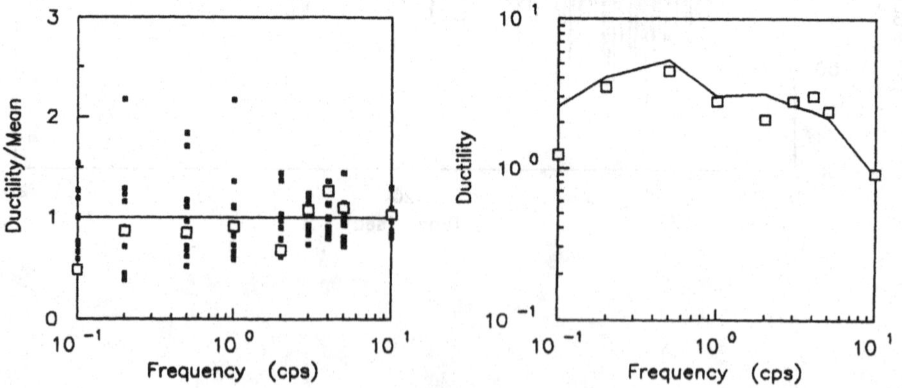

(c) Ventura, 1971

Fig. 5.2 Ductilities of a SDOF inelastic system under actual and simulated ground excitations

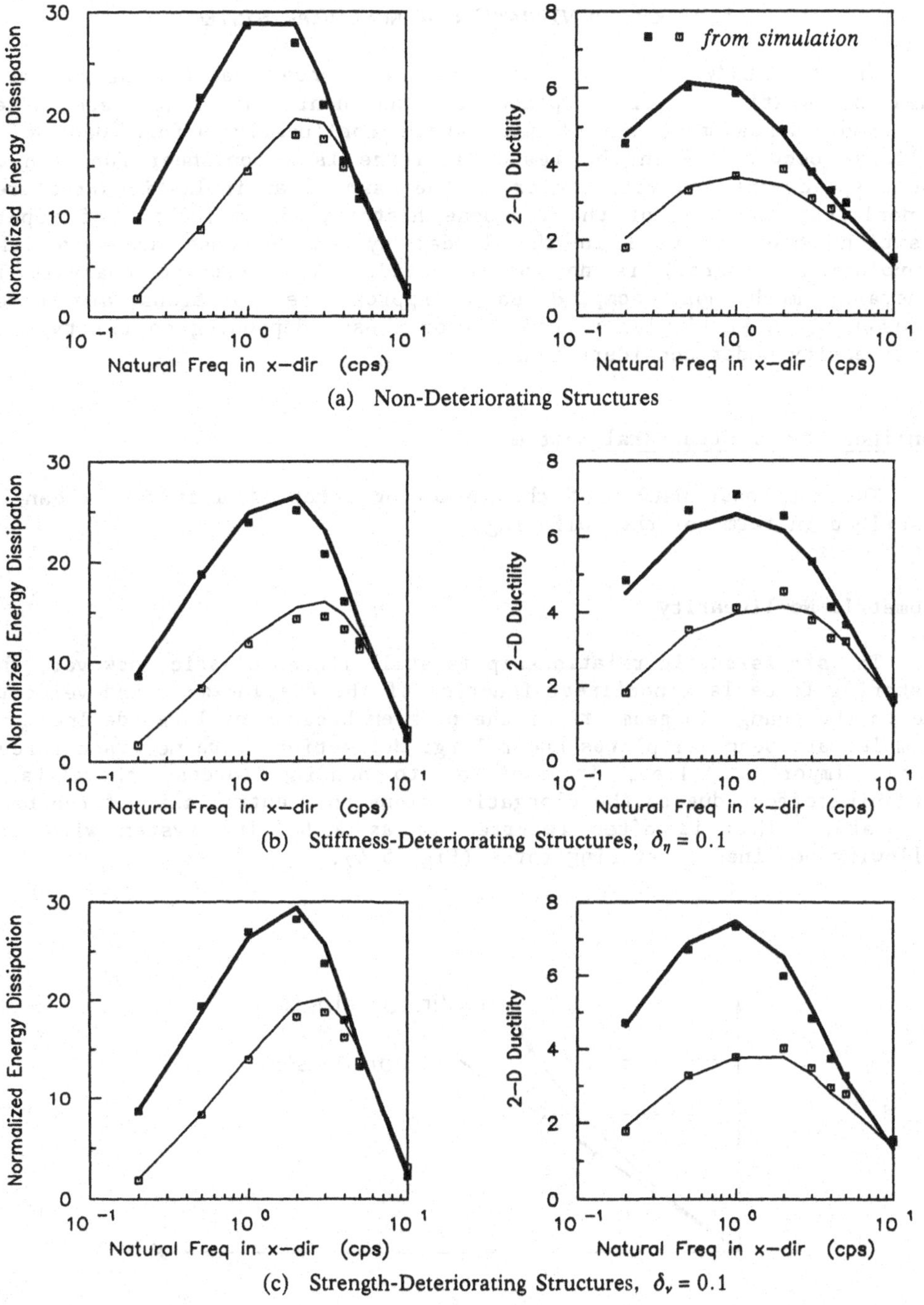

Fig. 5.3 Energy dissipations and ductilities of single-mass inelastic systems under non-stationary ground excitation; ----- time-variant frequency contents, ----- time-invariant frequency content; $k_x/k_y = 4$, $\Delta_x/\Delta_y = 0.5$.

5.2 RANDOM VIBRATION OF NONLINEAR SYSTEM

In the study of safety of structures under natural or man-made hazards, statistics of response in the nonlinear range are often required, since most structures behave nonlinearly before damage or collapse occurs. When the restoring force is a nonlinear function of the displacement and velocity or in the case of an inelastic structure, a nonlinear function of the response history, the principle of superposition which is used in the linear system response analysis (the convolution integral) is no longer valid. The response analysis is generally much more complicated. Approximate solutions are often necessary. A particular method may be chosen depending on the type of nonlinearity under considerations.

Nonlinearity of Structural System

The nonlinear nature of the restoring force of a structure can be described in terms of the following.

Geometric Nonlinearity

The stress-strain relationship is still linear elastic, however, the restoring force is a nonlinear function of the displacement and velocity due to the change in geometry of the problem because of large deflection. Examples are beams or plates under large deflection where membrane forces become important, i.e., in addition to bending forces, there is a hardening effect due to the elongation along the neutral axis of the beam or plate. This is often referred to as a Duffing system with the following nonlinear restoring force (Fig. 5.4).

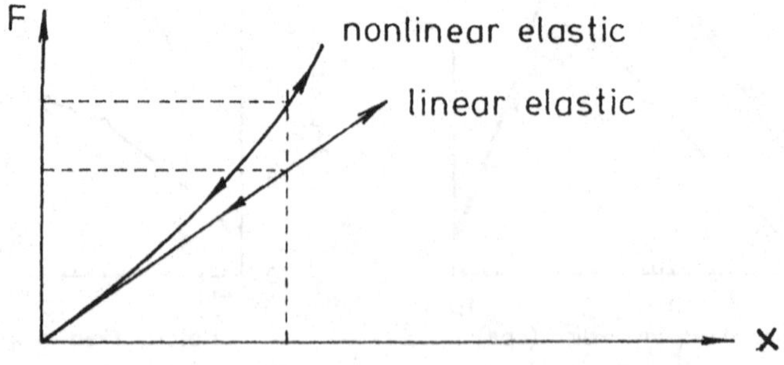

Figure 5.4 Duffing Hardening System

Another example is the snap-through problem associated with arches.

$$F = K_x[1 + \varepsilon \frac{x^2}{x_o^2}]$$

where the geometry change of the structure under loading causes significant loss in the system stiffness (i.e., a softening system, Fig. 5.5).

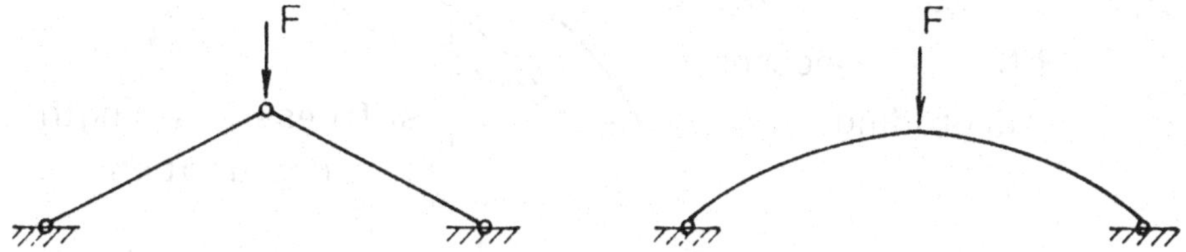

Figure 5.5 Softening Systems

Material Nonlinearity

The stress-strain relationship is no longer linear elastic, i.e., the restoring force cannot be expressed as an algebraic function of the displacement and velocity; it depends on the time history of the responses. The restoring force is inelastic and hysteretic. Most buildings and structures show hysteretic response behavior before damage or collapse occurs. For example, Fig. 5.6 shows the inelastic restoring force of a steel frame under large displacement.

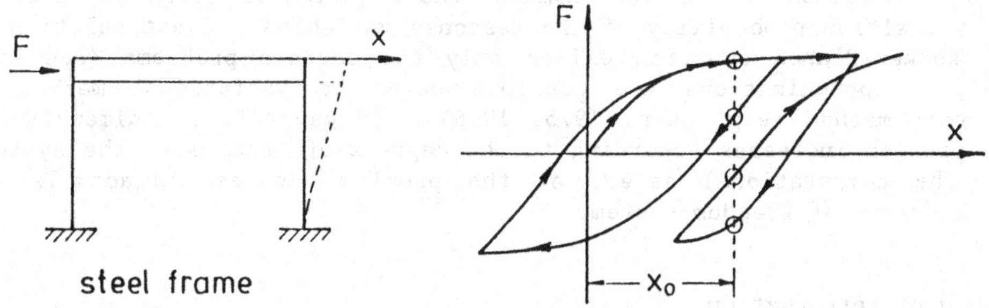

Figure 5.6 Steel Frame as a Hysteretic System

It is seen that there are many possible values of restoring force for a iven displacement. Only when the time history of the displacement is known can one determine the restoring force.

Figure 5.7 shows the restoring force of reinforced concrete or masonry buildings, the strength or stiffness or both of the system may deteriorate as the frame undergoes displacement reversals.

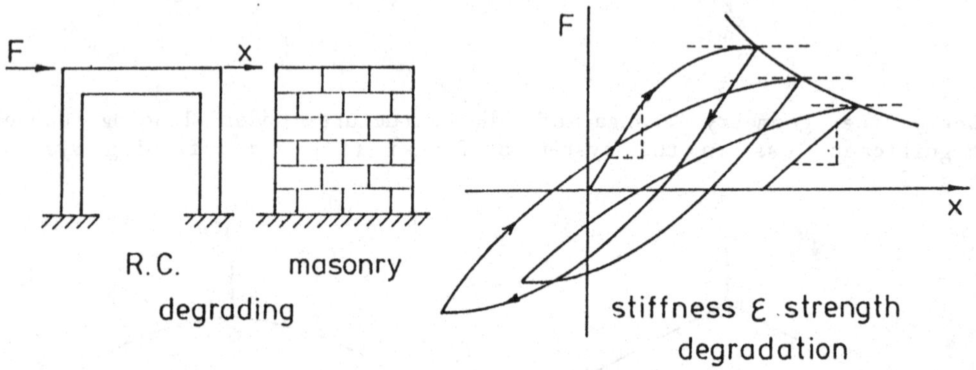

Figure 5.7 R.C. and Masonry Structures as Degrading Hysteretic Systems

Methods of Analysis of Nonlinear Structures Under Random Excitation

Exact solution of response of nonlinear system under random excitation exists only for some special cases under very restrictive conditions. For most nonlinear systems, only approximate solutions have been found. The different approaches are:

Markov-Vector and Fokker-Planck Equation

An exact formulation of the problem is possible using the Fokker-Planck equation of the joint density function of the response vector. It is a formulation in the time domain, the solution is given in terms of the transition probability of the response variables. Exact solution to the Fokker-Planck equation exists only for certain problems (see Lin, 1967). Approximations are possible based on variational method or Galerkin method (e.g., Wen, 1975, 1976). In general, the dimension of the problem increases according to the degrees-of-freedom of the system, and the computational aspect of the problem becomes intractable for multi-degree-of-freedom systems.

Method of Perturbation

It is a recursive scheme based on a perturbation of the associated linear system. The solution of the nonlinear system is replaced by a series of solutions of linear systems, applicable only for systems with small nonlinearity (e.g., Lin, 1967).

Method of Equivalent Linearization

The nonlinear system is replaced by an equivalent linear system in such a way that the mean square error (difference between the equations of motion of the two systems is minimized. There is no limitation of small nonlinearity, however, and iterations are generally required in the solution procedure.

In the following, because of its wide range of application and generally good accuracy, only the method of equivalent linearization will be covered. It was first applied to random vibration analysis by Booton (1954) and Caughey (1963), and since has been extended and applied to various nonlinear systems including hysteretic systems.

Consider a nonlinear system described by

$$\ddot{x} + g(\dot{x}, x) = F(t) \tag{5.21}$$

We seek an equivalent linear system with the following equation of motion

$$\ddot{x} + \beta_e \dot{x} + k_e x = F(t) \tag{5.22}$$

From the requirement of the minimization of the mean square value of the error

$$\varepsilon = g(\dot{x}, x) - \beta_e \dot{x} - k_e x \tag{5.23}$$

i.e., the equivalent damping coefficient and stiffness β_e and k_e are determined from

$$\begin{aligned} \frac{\partial}{\partial \beta_e} E[\varepsilon^2] &= 0 \\ \frac{\partial}{\partial k_e} E[\varepsilon^2] &= 0 \end{aligned} \tag{5.24}$$

The solution to Eq. 5.24, after substituting Eqs. 5.21 and 5.22 into 5.23, is

$$\beta_e = \frac{\sigma_x^2 E[\dot{x}g] - E[x\dot{x}]\, E[xg]}{\sigma_x^2 \sigma_{\dot{x}}^2 - E[x\dot{x}]^2}$$

$$k_e = \frac{\sigma_{\dot{x}}^2 E[xg] - E[x\dot{x}]\, E[\dot{x}g]}{\sigma_x^2 \sigma_{\dot{x}}^2 - E[x\dot{x}]^2} \tag{5.25}$$

If x(t) is stationary, $E[x\dot{x}] = 0$; Eq. 5.25 reduces to

$$\beta_e = \frac{E[\dot{x}g]}{\sigma_{\dot{x}}^2}, \qquad k_e = \frac{E[xg]}{\sigma_x^2} \tag{5.26}$$

Note that β_e and k_e are functions of the response statistics which are yet to be determined, therefore, the method generally leads to an iterative solution procedure. An example is given in the following on a Duffing oscillator under the excitation modeled by a Gaussian white noise.

The restoring force of a Duffing system is

$$g(x; x) = \beta\dot{x} + \omega^2(x + \alpha x^3) \tag{5.27}$$

$$E[\dot{x}g] = E[\beta\dot{x}^2 + \omega_o^2 x\dot{x} + \omega_o^2 x^3\dot{x}] \tag{5.28}$$

To evaluate the expected value, the joint density function of x and $\dot{x}$ is required, which in general is unknown and non-Gaussian even the excitation is a Gaussian process. As an approximation, one can assume that x and $\dot{x}$ are jointly Gaussian which allows one to obtain the following closed from solution,

$$E[\dot{x}g] = \beta\sigma_{\dot{x}}^2$$

$$E[xg] = E[\beta xx + \omega_o^2 x^2 + \alpha\omega_o^2 x^4]$$

$$= \omega_o^2\sigma_x^2 + \alpha\omega_o^2 E[x^4]$$

$$= \omega_o^2(\sigma_x^2 + 3\alpha\sigma_x^4) \tag{5.29}$$

From Eq. 5.26, one obtains

$$\beta_e = \beta \quad , \qquad k_e = \omega_o^2(1 + 3\alpha\sigma_x^2) \tag{5.30}$$

The stationary mean square response is, therefore,

$$\sigma_x^2 = \frac{\pi S_o}{2\beta_e k_e} = \frac{\pi S_o}{2\beta_e \omega_o^2(1 + 3\alpha\sigma_x^2)} \tag{5.31}$$

Equation 5.31 is a nonlinear algebraic equation for σ_x^2. Solving the quadratic equation, one obtains

$$\sigma_x^2 = \frac{1}{6\alpha} [(1 + 12\alpha\sigma_o^2)^{1/2} - 1]$$

$$\approx \sigma_o^2(1 - 3\alpha\sigma_o^2) \qquad \text{for small } \alpha \tag{5.32}$$

in which $\sigma_o^2 = \pi S_o/2\beta\omega_o^2$, the mean square response of the corresponding linear system (i.e., α=0). It is seen that the mean square response of the nonlinear system decreases as α increases, typical behavior of a hardening system.

The equivalent linearization method can be extended to M.D.F. system without difficulty (Iwan, 1973). However, the linearized system coefficients often are in an implicit form which are not very useful. A method by Atalik and Utku (1976) gives explicit solutions for the parameters of the linearized system under some general conditions which proves most convenient for M.D.F. systems.

The equation of motion of a M.D.F. nonlinear system may be given by

$$g(\ddot{x}, \dot{x}, x) = f(t) \tag{5.33}$$

in which g = a vector representing the total internal force acting in the ith degree-of-freedom system, a single value, odd function,
f = stationary Gaussian excitation vector with zero mean.

Equation 5.33 is replaced by a linear system

$$\mathbf{M}\,\ddot{\mathbf{x}} + \mathbf{c}\,\dot{\mathbf{x}} + \mathbf{k}\,\mathbf{x} = \mathbf{f}(t) \tag{5.34}$$

in which the mass, damping and stiffness-matrices are determined by the requirement that the mean square error $E[\epsilon^T \epsilon]$ is minimized where T denotes transpose and where $\epsilon = g - M\ddot{x} - c\dot{x} - kx$. The necessary condition for minimization is

$$\frac{\partial E[\epsilon^T \epsilon]}{\partial m_{ij}} = 0 , \qquad \frac{\partial E[\epsilon^T \epsilon]}{\partial c_{ij}} = 0 , \qquad \frac{\partial E[\epsilon^T \epsilon]}{\partial k_{ij}} = 0 \tag{5.35}$$

Equation 5.36 leads to the following solution

$$m_{ij} = E\left[\frac{\partial g_i}{\partial \ddot{x}_j}\right] , \qquad c_{ij} = E\left[\frac{\partial g_i}{\partial \dot{x}_j}\right] , \qquad k_{ij} = \left[\frac{\partial g_i}{\partial x_j}\right] \tag{5.36}$$

in which g_i = ith element of g and x_j = jth element of **x**. The sufficient condition proof of Eq. 5.36 was given by Atalik and Utku (1976). Necessary condition proof can be found in Chen (1987) and Faravelli, et al. (1988).

From Eq. 5.36, one can evaluate the required parameters of the linearized system. To solve for the response variable statistics, one can follow the general procedure

(1) Start the solution by assuming the system to be linear. Calculate the response variable covariance matrix and other response statistics that are required in Eq. 5.36 and calculate the system parameters.

(2) Solve Eq. 5.34 for new response statistics.

(3) Calculate the new system parameters using response statistics obtained in Step 2 in Eq. 5.36.

(4) Repeat Steps 2 and 3 until the solution converges.

Hysteretic Systems

Since the restoring force in a hysteretic system depends on the time history of the displacement, proper modeling of the restoring force is necessary. In deterministic response analysis, often empirical rules of hysteresis are used, e.g., the widely used elasto-plastic and bilinear systems. Even with such simplifying rules, terms $E[\dot{x}g]$ and $E[xg]$, etc., in Eq. 6 are still difficult to evaluate because of the hereditary nature of the restoring force. Caughey (1963), using a technique of slowly varying parameter (or Krylov-Bogoliubov) technique, i.e., assuming

$$x(t) = R(t)\cos[\omega_e t + \theta(t)] \tag{5.37}$$

in which R(t) and θ(t) are response amplitude and phase angle, slowly varying functions of time. By further assuming that the peak response follows a Rayleigh distribution, he obtained the equivalent damping coefficients and stiffness of a bilinear system (see Fig. 5.8)

$$\beta_e = \beta_o + \frac{2}{\sqrt{\pi}}\frac{1-\alpha}{k_e}\frac{Y}{\sqrt{2}\sigma_x}\,\mathrm{erf}\left[\frac{Y}{\sqrt{2}\sigma_x}\right]$$

$$k_e = \frac{F_Y}{Y}\left\{1 - \frac{8(1-\alpha)}{\pi}\int_1^\infty [(z\lambda)^{-1}(z-1)^2 + z^{-3}(z-1)^{1/2}]e^{-z^2/\lambda}dz\right] \tag{5.38}$$

in which β_o = viscous damping coefficient
α = post pre-yield stiffness ratio
Y = yielding displacement
F_Y = yielding force
$\lambda = 2\sigma_x^2/Y^2$.

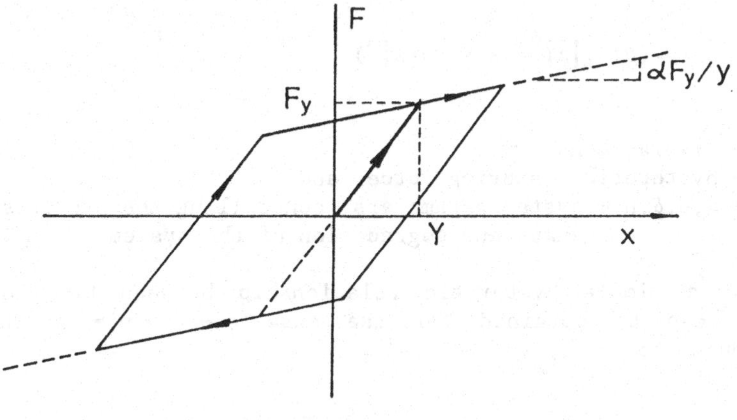

Figure 5.8 Bi-linear Hysteretic Restoring Force

From Eq. 5.38 one obtains the mean square response as

$$\sigma_x^2 = \frac{\pi S_o}{2\beta_e k_e} \tag{5.39}$$

Again, an iterative procedure is required since both β_e and k_e are functions of σ_x. Comparison of results with Monte Carlo simulation (Iwan, 1968) indicates that when the nonlinearity is small ($\alpha > 0.5$), the accuracy is satisfactory. However, for a nearly elasto-plastic system ($\alpha < 0.1$), the accuracy is rather poor in the range $0.5 < \sigma_x/Y < 5$, the error can be so much as 50-60% on the unconservative side (underestimated). This inaccuracy, in fact, is more due to the slowly-varying-parameter assumption than the linearization procedure, since Eq. 5.37 excludes any possibility of "drifting" which is common for inelastic system, and is essentially a narrow-band assumption, and as a result, the damping is overestimated, causing underestimations in the response statistics.

An alternative method is recently proposed by Wen (1980) and Baber and Wen (1981). In this method, no empirical rule of hysteresis is used, instead the restoring force is modeled by a nonlinear differential equation which takes the time-history dependent nature of the forcing function into consideration. No other assumptions (such as the Krylov-Bogoliubov technique) are required. The solution, as a result, can be obtained in closed form and has better accuracy. Also, it can be extended to M.D.F. and degrading systems without difficulty.

Consider a S.D.F. system, the hysteretic restoring force is modeled by a first order, nonlinear differential equation

$$\dot{z} = \frac{1}{\eta}\,[A\dot{u} - \nu(\beta|\dot{u}||z|^{n-1}z + \gamma u|z|^{n})] \tag{5.40}$$

in which u = displacement,
z = hysteretic restoring force, and
$\eta, A, \nu, \beta, \delta, n$ = system parameters, controlling the hysteresis shape and degradation of the system.

For example, a simple hysteretic relationship between the force and displacement can be obtained for the case $\eta = \nu = n = 1$, that is, Eq. 5.40 reduces to

$$\dot{z} = A\dot{u} - \beta|\dot{u}|z - \gamma\dot{u}|z| \tag{5.41}$$

According to whether $\dot{u}$ and z are positive or negative, Eq. 5.41 can be rewritten in Table 5.1.

Table 5.1 Change of Stiffness in Loading and Unloading

$\frac{dz}{du} = A - (\gamma-\beta)z$	$\frac{dz}{du} = A - (\beta+\gamma)z$	
(2)	(1)	$\dot{u}$
(3)	(4)	
$\frac{dz}{du} = A + (\beta+\gamma)z$	$\frac{dz}{du} = A - (\beta-\gamma)z$	

For example, for a displacement u which starts from zero and undergoes a cyclic motion, the restoring force would follow the four equations in the order indicated in Table 5.1 and the force-displacement relationship is given in Fig. 5.9 for the case $\gamma = \beta = 0.5$.

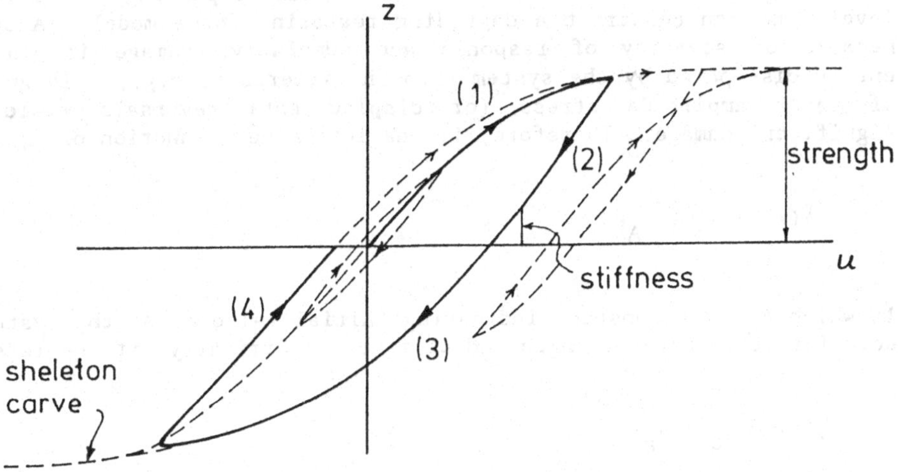

Figure 5.9 Restoring Force and Displacement Relationship

By adjusting γ and β, one can obtain a large class of hysteretic hardening as well as softening systems with various degrees of energy dissipation capacity. Value of n controls the smoothness of the transition from elastic to inelastic range. For example, for large n, the hysteresis approaches that of an elasto-plastic system. Some of these properties of the restoring force are illustrated in Figs. 5.10 through 5.12.

It can be shown (Baber and Wen, 1979) that the strength of the system (maximum restoring force) is given by

$$\left[\frac{A}{\nu(\beta+\gamma)}\right]^{1/n} \tag{5.42}$$

the stiffness at z=0 is

$$\left[\frac{dz}{du}\right]_{z=0} = A/\eta \tag{5.43}$$

and the energy dissipation in cyclic motion is

$$\varepsilon = \int_0^T z\dot{u}\, dt \tag{5.44}$$

If one lets the parameters be functions of time depending on the response level, one can construct a degrading restoring force model. A suitable measure of severity of response and cumulative damage is the total energy dissipated by the system through hysteresis, e.g., a large number of large amplitude stress (or displacement) reversals would cause significant damage. Therefore, if one lets A be a function of ε, e.g.,

$$A(\varepsilon) = A_o - \delta_A \varepsilon \tag{5.45}$$

in which A_o is a constant indicating initial value of A, the system will deteriorate in both strength and stiffness. Similarly, if one lets only

$$\eta(\varepsilon) = \eta_o + \delta_\eta \varepsilon \tag{5.46}$$

the system will have only stiffness deterioration, whereas if only

$$\nu(\varepsilon) = \nu_o + \delta_\nu \varepsilon \tag{5.47}$$

the system will deteriorate in strength only. δ_A, δ_η, and δ_ν are constants controlling the rate of deterioration.

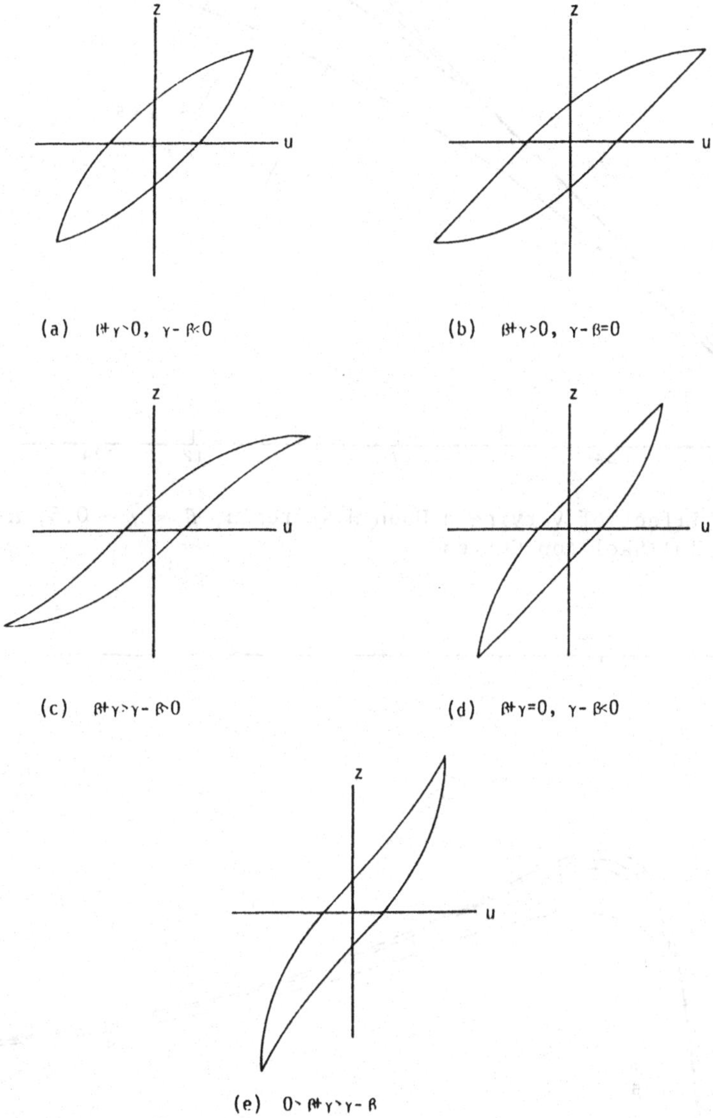

Figure 5.10 Possible Combinations of β and γ

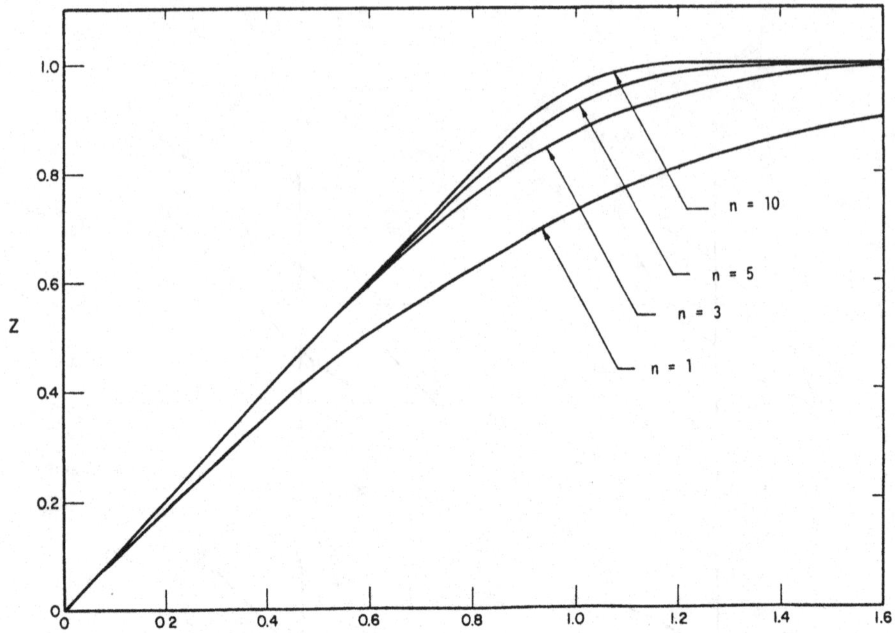

Figure 5.11 Effect of Varying n Upon Hysteresis, $\beta = \gamma = 0.5$, A=1
(a) Skeleton Curves

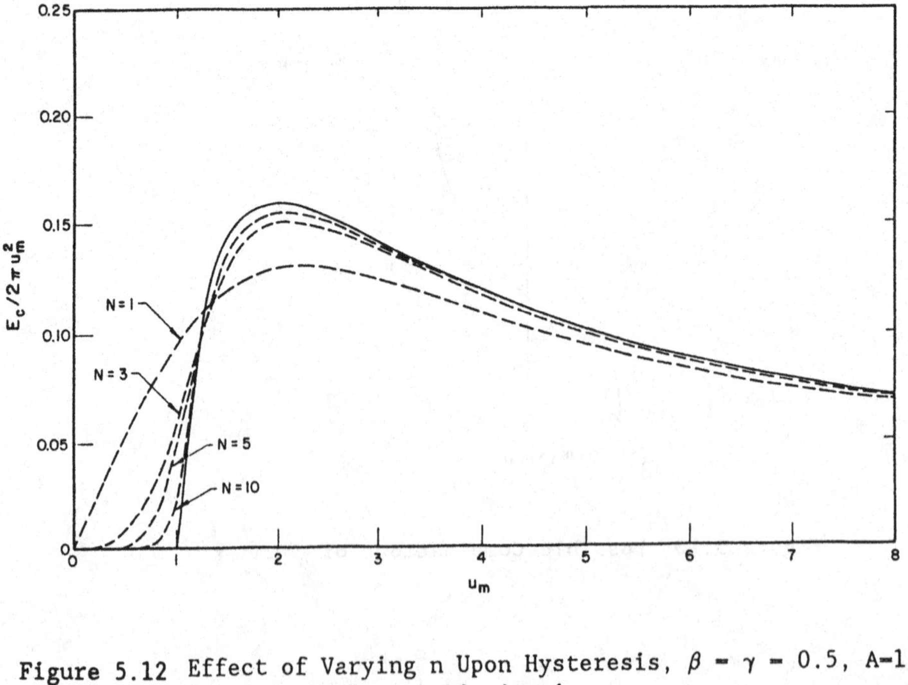

Figure 5.12 Effect of Varying n Upon Hysteresis, $\beta = \gamma = 0.5$, A=1
(b) Cyclic Energy Dissipation

Inelastic systems using Eq. 5.40 for the hysteretic restoring force under random excitation can be linearized in closed form based on the Atalik and Utku procedure, and both stationary and nonstationary solutions can be obtained. Details are given in Wen (1980) and Baber and Wen (1981). Monte Carlo simulation results indicate that the accuracy of linearization using the Atalik and Utku method is good for all response levels for nondegrading as well degrading systems. Also extension to M.D.F. systems, biaxial systems, and torsional motion studies are quite satisfactory. A review of this method and other approximate solutions is given recently in Wen (1989).

The response and safety analysis of structures with such hysteretic restoring force under random excitation and applications to earthquake engineering problems can be found in Baber and Wen (1981), Baber and Wen (1979), Sues, Wen, and Ang (1985), and Park, Ang, and Wen (1985).

5.3 RESPONSE AND DAMAGE ANALYSIS OF INELASTIC STRUCTURES

Since the late fifties, research on random vibration of structures and structural systems has made great progress. Some of the research results have been successfully applied in design, e.g., the gust response factor approach used in most modern building codes for wind load. Most of the applications, however, have been limited to the case that the structural response stays within the linear-elastic range. The obvious reason is that for such systems the structural restoring force is easy to model and the response statistics can be obtained by well-developed linear random vibration methods.

To study the performance and safety of structures under severe loads, e.g., against damage and collapse, however, the inelastic response behavior needs to be taken into consideration. As is shown in the foregoing that when the structure becomes inelastic, the restoring force is highly nonlinear and hysteretic. It depends on the time history of the response and may deteriorate in strength or stiffness, or both, when the oscillation progresses for an extended period of time. The modeling of the restoring force has been treated in the previous section. The response analysis of the inelastic structure under random excitation and the associated method of predicting damage are briefly outlines in this chapter.

Biaxial Interaction

For two-dimensional structures under biaxial excitations, the interaction of the restoring forces in the two directions may significantly alter the response behavior. For example, the damage suffered from oscillations in one direction is likely to weaken the strength and/or stiffness in the other direction and vice versa. The restoring force model given in the previous chapter has been extended to include such interaction by requiring that the hysteretic components in the two directions, i.e., z_x and z_y, satisfy the following coupled differential equations (Park, et al., 1986)

$$\dot{z}_x = A\,\dot{u}_x - \beta|\dot{u}_x z_x| z_x - \gamma\,\dot{u}_x z_x^2 - \beta|\dot{u}_y z_y| z_x - \gamma\,\dot{u}_y z_x z_y$$

$$\dot{z}_y = A\,\dot{u}_y - \beta|\dot{u}_y z_y| z_y - \gamma\,\dot{u}_y z_y^2 - \beta|\dot{u}_x z_x| z_y - \gamma\,\dot{u}_x z_x z_y \qquad (5.48)$$

in which u_x and u_y are, respectively, the displacements in the x and y directions. A, β, and γ are system parameters. The above equations give an isotropic hysteretic restoring force. It can be easily demonstrated that both equations reduce to the case of uniaxial loading along an arbitrary direction. For an orthotropic system whose stiffness and strength in the two directions are different, one can introduce a simple transformation (scaling) of the response variables and still use the same equations. As in the uniaxial model, deterioration can be introduced by letting parameters A, β, and γ be functions of time depending on the severity of the response; e.g., maximum response amplitude or hysteretic energy dissipation given by

$$\varepsilon_T(t) = \int_0^t [z_x(\tau)\dot{u}_x(\tau) + z_y(\tau)\dot{u}_y(\tau)]d\tau \qquad (5.49)$$

or both.

The accuracy and capability of this method of modeling is indicated by comparisons of the force-displacement relationship with those based on plasticity theory and experimental studies. Figures 5.13 and 5.14 show the nondegrading system according to Eq. 5.48 under different displacement paths and the corresponding analytical solutions (Powell and Chen, 1986) based on plasticity theory. The agreements are surprisingly good considering the generally complicated biaxial inelastic stress strain relationship and the simple and explicit nature of the model.

Method of Solution Under Random Excitation

The hereditary behavior of the restoring force of the inelastic system makes analytical solution of the response extremely difficult. Approximations are almost always necessary.

As mentioned in the foregoing, the equivalent linearization method has been used by Caughey for bilinear systems under white noise excitation. An excellent literature survey of the method was given in (Spanos, 1981). For the hysteretic system as indicated in the foregoing a major difficulty is to perform the linearization in closed form because of the hereditary behavior of the restoring force. The smooth hysteretic model, however, allows such linearization to be performed in closed form without recourse to the K-B technique (Wen, 1980; Baber and Wen, 1981). This greatly facilitates the applicability of the linearization method to multi-degree-of-freedom systems and degrading systems. For example, for a nondegrading S.D.O.F. system with a restoring force given by Eq. 21 of the previous chapter ($n=\eta=\nu=1$) under zero mean, Gaussian excitation, by assuming the response variables to be also jointed Gaussian, the equation can be linearized as

$$\dot{z} + c_1\dot{x} + c_2 z = 0 \qquad (5.50)$$

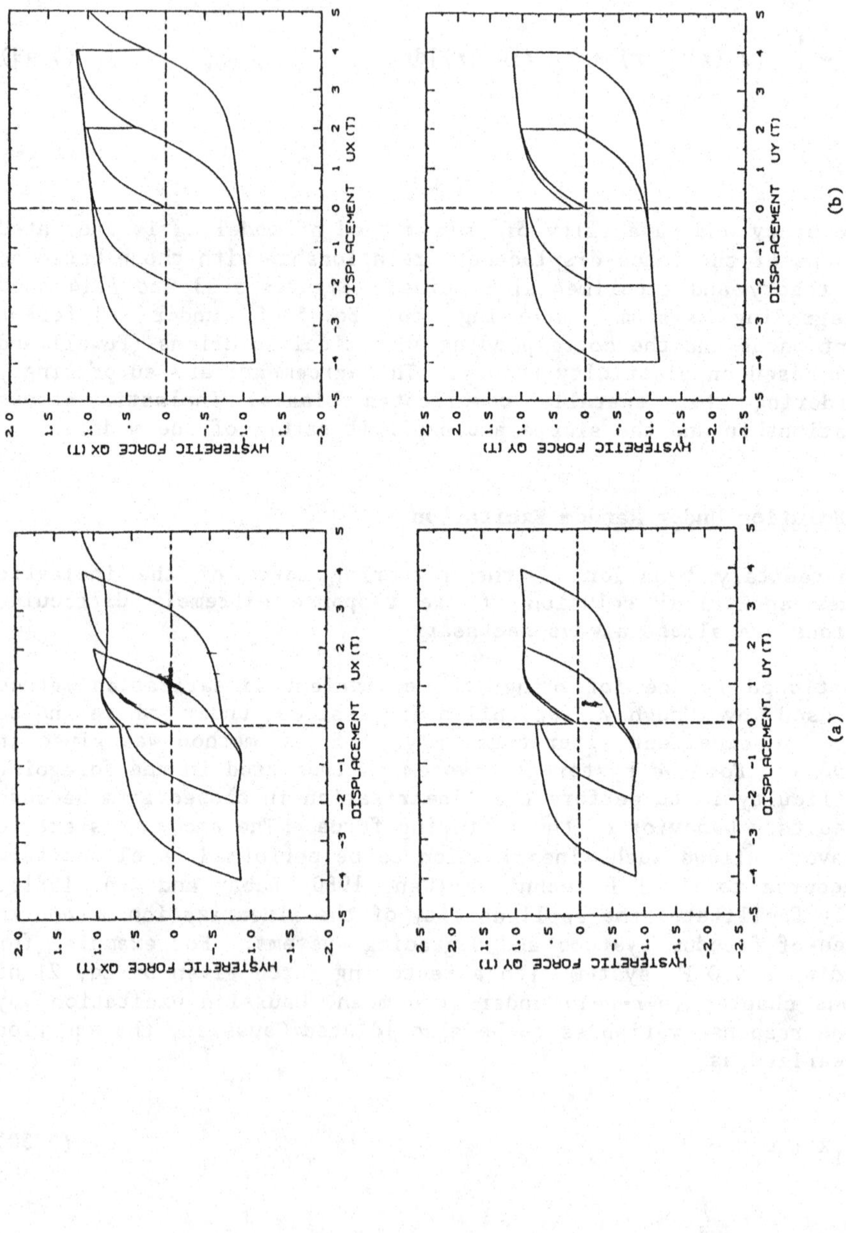

Fig. 5.13 Nondegrading Biaxial Smooth Hysteretic Restoring Force:
(a) Diamond Displacement Path; (b) Square Displacement Path

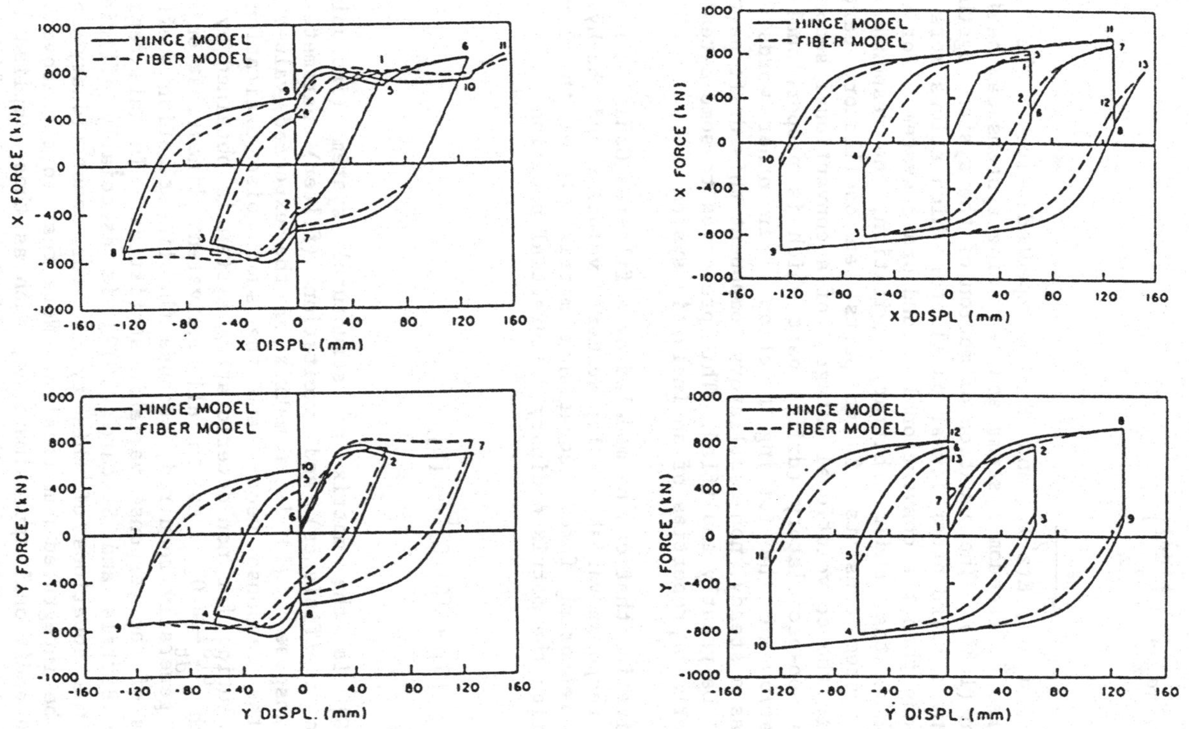

Fig. 5.14 Force-Displacement Relationship of Analytical Model by Powell and Chen (1986)

in which,

$$c_1 = \sqrt{\frac{2}{\pi}} \, [\gamma \, \frac{E(\dot{x}z)}{\sigma_{\dot{x}}} + \beta\sigma_z] - A \tag{5.51}$$

$$c_2 = \sqrt{\frac{2}{\pi}} \, [\gamma\sigma_{\dot{x}} + \beta \, \frac{E(\dot{x}\dot{z})}{\sigma_z}] \tag{5.52}$$

$\sigma_{\dot{x}} = \sqrt{E(\dot{x}^2)}$, $\sigma_z = \sqrt{E(z^2)}$, and E[ ] = expected value. The extension to multi-degree of freedom systems and degrading systems can be found in Baber and Wen (1981). The equation of motion of the system together with Eq. 5.50 forms a third-order linear oscillator which allows more freedom in the response than the conventional second-order system. For example, it can be shown that under arbitrary excitation, the response of this third-order system consists of two parts; an oscillatory part which corresponds to the convolutional integral of a conventional second-order system and a non-oscillatory (drift) part which is proportional to the time integration of the forcing function. In other words, if the excitation has a steady, non-oscillatory component this drift part of the response will be greatly amplified. This preserves to some extent one of the most important properties of an inelastic system.

For excitation that can be modeled by a filtered Gaussian shot noise the one-time response variable (displacement, velocity and the hysteretic part of the restoring force) covariance matrix [S] of the linearized system satisfies the matrix ordinary differential equation

$$\frac{d[S]}{dt} + [G][S] + [S][G]^t = [B] \tag{5.53}$$

in which [G] is the matrix of structural system (including the linearization coefficients) and excitation (filter) parameters. t indicates transpose. [B] is a matrix of the expected values of the product of the response vector and the shot noise excitation. The stationary solution for non-deteriorating system is obtained by solving Eq. 5.53 with $\frac{d[S]}{dt} = 0$. As in all equivalent linearization method, iteration is generally required, because the linearization coefficients are functions of the response variable statistics. In this regard, the algorithm in Bartels and Stweard (1972) is especially efficient for solution of the equations of this type. For nonstationary solution, Eq. 5.53 can be integrated numerically. Note that in the above solution procedure, no additional approximations, such as those implied in a K-B technique, have been made other than that in the equivalent linearization and the Gaussian assumption for the response variables. Therefore, the solution so obtained is the best linear approximation of the original nonlinear system under the circumstances.

For degrading systems, to keep the analysis tractable, an additional approximation is necessary, i.e., A, η, and ν are assumed to be slowly varying and can be approximated by their mean values, μ_A, μ_η, and μ_ν. Therefore, taking expected values of the time derivatives, one obtains the time derivative of μ_A as a function of expected energy dissipation rate, i.e., $(1-\alpha)kE[\dot{x}z]$. Since $E(\dot{x}z)$ is part of the solution [S], μ_A, can be updated at each time step to reflect the system deterioration. Similarly one can update μ_η and μ_ν. The above approximation has been verified by extensive Monte Carlo simulations (Baber and Wen, 1981).

The power spectral density matrix of the response variables (in the case of stationary response) can be obtained through an eigen analysis of the matrix [G]. For a multi-D.O.F. system, the response power spectral density matrix is given by Baber and Wen (1981)

$$W_{yy}(\omega) = \Phi[i\omega I-\Sigma]^{-1}\psi^t W_{ff}\psi^*[i\omega I-\Sigma]^{-1^*}\Phi^{*t} \tag{5.54}$$

in which W_{ff} = constant excitation power spectral density matrix, Φ and ψ are the left and right eigenvector matrices of [G]. I = identity matrix and Σ = the diagonal matrix of eigenvalues, t indicates transpose and * indicates complex conjugate. Note that [G] is not a symmetric matrix. Although a frequency domain approach is possible through iterative solutions of Eqs. 5.53 and 5.54, the computational effort can be extensive, since numerical integration is generally required in each iteration. Therefore, a time domain solution for the system coefficients first is preferred if the power spectral density is needed. The statistics of maximum response can be obtained by using currently available approximate procedures, e.g., that based on a Poisson outcrossing assumption. A response quantity that is particularly useful for predicting potential structural damage is the total hysteretic energy dissipation $\varepsilon_T(t)$. The mean value of $\varepsilon_T(t)$ is

$$E[\varepsilon_T(t)] = (1-\alpha)k\int_0^t E[z(\tau)\dot{x}(\tau)]d\tau \tag{5.55}$$

As $E(z\dot{x})$ is part of the solution of [S], $E[\varepsilon_T(t)]$ can be easily evaluated. Evaluation of the variance of $\varepsilon_T(t)$ requires the solution of the covariance matrix between two time instants. Details of the solution procedure can be found in Pires, et al. (1983).

The method of solution outlined has been also extended to the systems under biaxial and torsional excitation, and 3-D frames (Wen and Yeh, 1989; Casciati and Faravelli, 1987).

The accuracy of the method has been found to be generally very good based on comparison with results from extensive computer simulations.

Figures 5.15 and 5.16 show the comparisons of some of the response variable statistics with those from computer simulations for both uniaxial and biaxial systems of single as well as multi-degree-of-freedom. As a result, this method has been used quite successfully in safety and performance study of building and structural systems under seismic excitation. Some details of the applications to damage prediction will be given in the following.

When the excitation spectral content is such that the power spectral density function vanishes rapidly as the frequency goes to zero, however, the method tends to underestimate the displacement response. The error largely depends on the characteristics of the system restoring force and the excitation in the low frequency range. This is primarily due to the fact indicated earlier that the linearized system picks up the drift due to the non-oscillatory component of the excitation only. It does not predict well the drift due to the high frequency component of the excitation (Iwan, et al., 1987). This error is negligible when low frequency component of the excitation exists and the drift response to this component dominates; for example, for systems under earthquake excitation with a Kanai-Tajimi type spectral density function which is nonzero at zero frequency. On the other hand, for an earthquake excitation with a power spectral density which goes to zero at zero frequency, for example, according to the square of the frequency at low frequencies, the r.m.s. response could be underestimated by 20 to 30 percent. Therefore, caution is necessary for excitations of such characteristics when using this method.

Damage Analysis

The application of the foregoing methodologies to actual engineering systems has been limited and largely restricted to response and damage studies of buildings and structures under strong motion earthquake excitation. In most studies the smooth hysteresis model and the method of equivalent linearization were used. The obvious reason is that actual buildings and structures generally are complex and need to be treated as multi-degree-of-freedom systems. Also, deterioration in such systems occurs as a rule rather than exception when under severe excitation such as those due to earthquakes. This particular method of analysis is, therefore, most suitable under such circumstances. Examples of applications of this method to damage prediction of reinforced concrete and masonry structures are given in the following.

Reinforced concrete buildings generally show hysteretic and degrading response behavior under severe earthquake excitation. Sues, et al. (1985) applied the methodology to a number of reinforced concrete buildings that were damaged during recent earthquakes. Since considerable uncertainty exists in the structural system itself, in addition to structural response, they also considered the uncertainty of the structural system parameters (such as stiffness, damping, hysteretic force parameters) and sensitivity of the response to changes in the structural parameters. They obtained the sensitivity coefficients (the

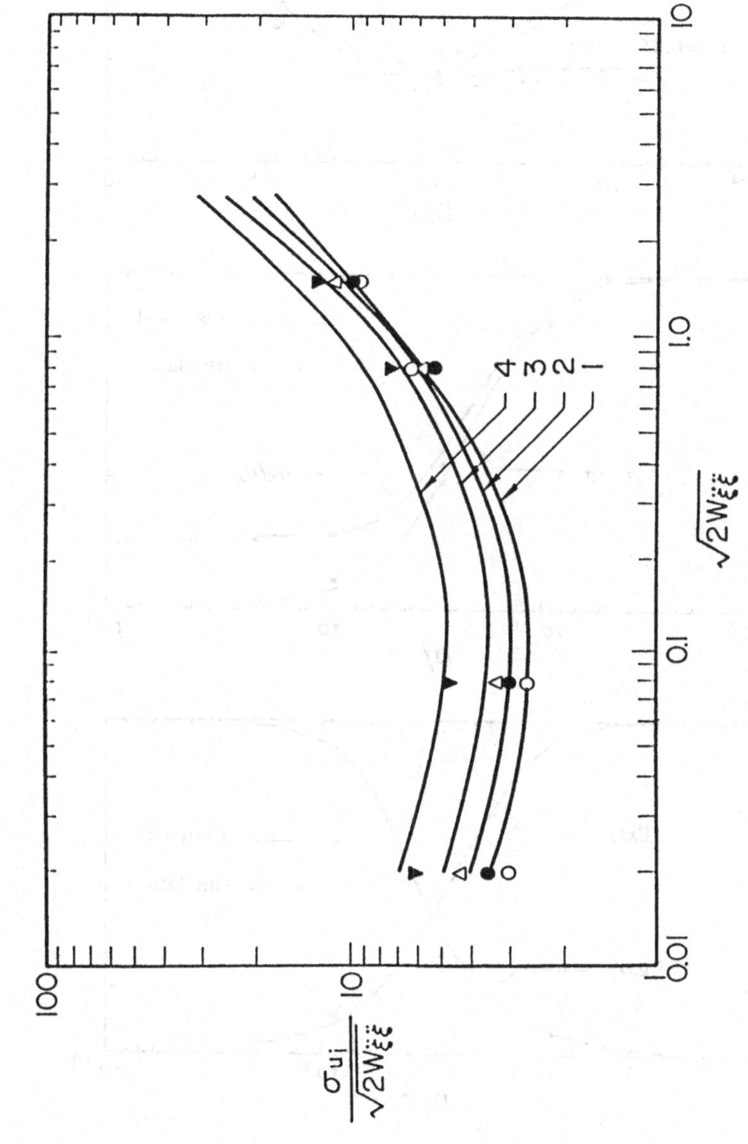

Fig. 5.15 Comparison of R.M.S. Interstory Displacement (u_i) of a Four-Story Building with Computer Simulation $W_{\ddot{\xi}\ddot{\xi}}$ Power Spectral Density of the White Noise Excitation

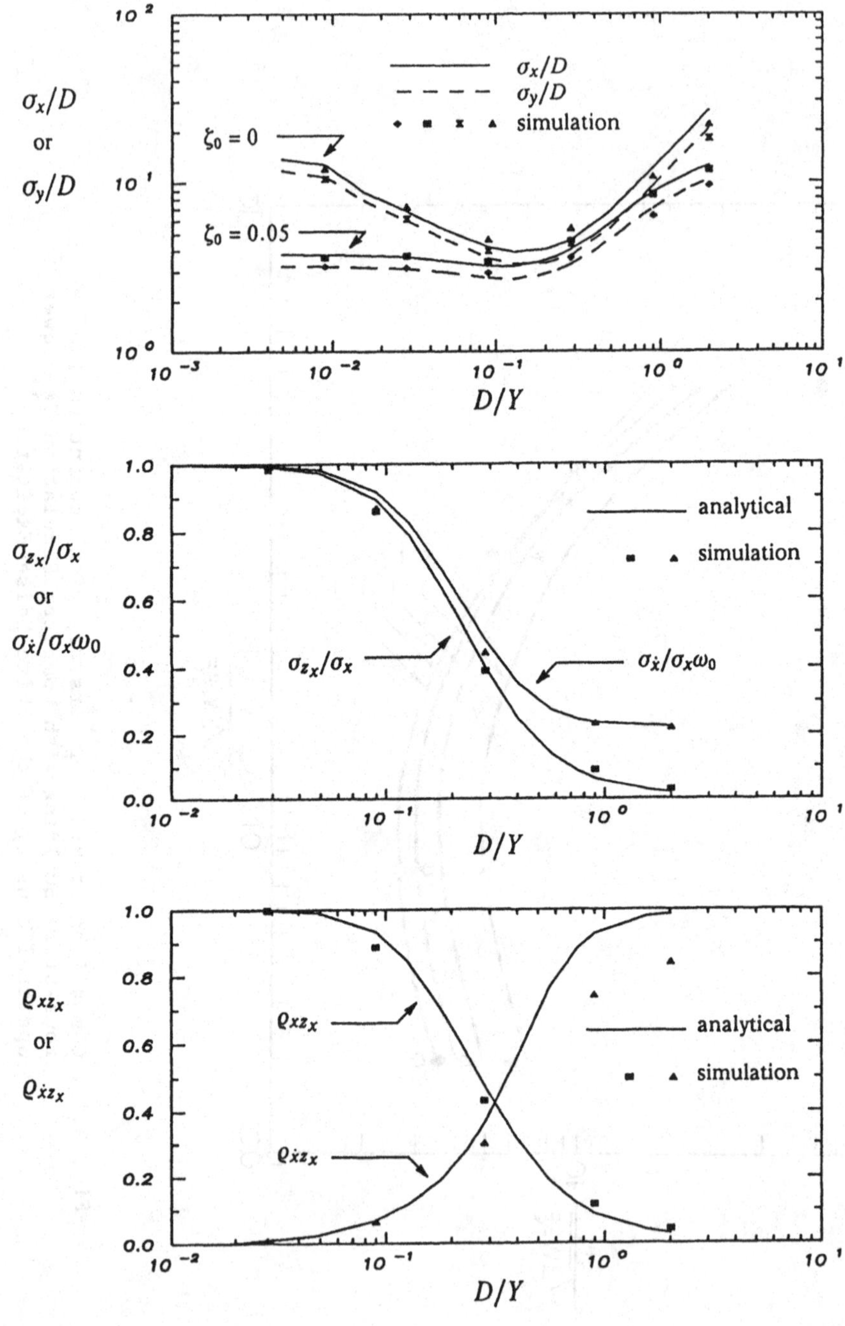

Fig. 5.16 Response of S.D.O.F. Biaxial Smooth System Under Bi-Directional Excitation. D - Excitation Intensity, Y - Yield Displacement, ρ - Correlation Coefficient, ξ_0 - Viscous Damping Ratio

rate of change of response statistics with the system parameter) required in the analysis by solving the following equation which can be derived from Eq. 5.53,

$$\frac{\partial}{\partial t}\left(\frac{\partial S}{\partial p}\right) + G\,\frac{\partial S}{\partial p} + \frac{\partial G}{\partial p}\,S + \frac{\partial S}{\partial p}\,G^T + S\,\frac{\partial G^T}{\partial p} = \frac{\partial B}{\partial p} \tag{5.56}$$

in which, $\frac{\partial S}{\partial p}$ is the unknown matrix of the sensitivity coefficient. The above equations are linear and can be solved easily. Note that typical sensitivity analyses require repeated solutions of the original nonlinear system.

The variance due to parameter uncertainty was then combined with that due to random vibration to arrive at the overall uncertainty and hence, probabilities of limit states being reached given the excitation intensity. For example, the maximum story drift was calculated for the Holiday Inn Building in Van Nuys, California, which is a seven-story reinforced concrete structure that sustained extensive nonstructural damage and moderate structural damage during the 1971 San Fernando Earthquake. The mean maximum drifts of the first four stories were calculated to be 1.59%, 1.63%, 1.27%, and 0.68% of story height; as compared with 0.9%, 1.14%, 1.48%, and 1.50% based on a linear time history analysis and actual recorded ground motion. The uncertainty analysis showed that given the excitation intensity, the coefficient of variation of the total response ranged between 60% and 80%; approximately 50~80% of which is contributed by the parameter uncertainties, whereas 10~30% from the random vibration and 10~20% from structural modeling uncertainty. This indicates the importance of the uncertainties in the structural parameters.

Since damage can be caused by repeated stress reversals as well as high stress excursion, the damage may be realistically expressed as a combination of the maximum deformation δ_m and the hysteretic energy dissipation, i.e., $\varepsilon_T(t)$ of Eq. 5.55. Such an index (Park, et al., 1985) for reinforced concrete structures is given in the following form,

$$D = \frac{\delta_m}{\delta_u} + \frac{\beta}{Q_y \delta_u}\,\varepsilon_T(t) \tag{5.57}$$

in which D = damage index (positive values represent different degrees of damage and $D \geq 1.0$ represents collapse); δ_u = ultimate deformation under static loadings; Q_y = yield strength; β = nonnegative constant. Q_y, δ_u, and β are parameters of structural (or member) capacity against damage and collapse. A large set of U.S. and Japanese test data of reinforced concrete beams and columns tested to failure were analyzed to determine the value of Q_y, δ_u, and β and to determine the uncertainty in the ultimate member capacity in terms of D. The overall damage index D_T of a structure can be evaluated from those of the members based on an energy consideration. It reflects the damage concentration at the weakest part of a building (e.g., the first story or top story as frequently observed as well as damage distribution throughout a building).

In terms of the damage index, D_T, nine reinforced concrete buildings that were moderately or severely damage during the 1971 San Fernando earthquake and the 1978 Miyahiken-Oki earthquake in Japan were evaluated by Park, et al. (1985). Figure 5.17 gives the plans and elevations of these buildings; earthquake excitation is assumed in the longitudinal direction. In calculating D_T by the random vibration method, the best estimated intensity and frequency content of the excitation at the site were used. Figure 5.18 shows the calculated D_T versus the observed damage, which provide a basis for defining the significance of values of D_T. It is found that the index value $D_t \leq 0.4$ approximately corresponds to reparable damage, whereas $D_T > 0.4$ represents damage beyond repair and $D_T > 1.0$ represents total collapse. A damage limiting design procedure based on the above consideration has also been proposed by Park, et al. (1987).

Following a similar procedure Kwok (1987) studied the damage of masonry buildings under earthquake excitation. Although masonry structures were known to be brittle, it was found that energy dissipation is also significant in the damage analysis. The damage analysis was applied to a total of 18 buildings in four cities in China which were damaged during the 1975 Haicheng earthquake. The comparison of the calculated damage index with the observed damage is shown in Fig. 5.19. It is seen that an index value of 0.3 corresponds approximately to the borderline between severe and moderate damage which may be used as allowable damage level. A damage limiting design procedure was also proposed.

5.4 FAST INTEGRATION METHOD

In evaluating structural reliability under stochastic loadings, the emphasis is normally on the uncertainties due to random time histories of the loadings, individually or in combination. Great progress has been made in random process and random vibration methods for modeling the loadings and evaluating the required response statistics to perform the reliability analysis. The structural system properties such as stiffness, damping and strength and the excitation parameters such as frequency content and duration (hereafter referred to as "system parameters"), usually assumed given in the stochastic analysis, however, are seldom perfectly known, or can be predicted with certainty. The effect of the system parameter uncertainties may be equally important, even dominant as far as the overall reliability of the structural system is concerned.

To include the parameter uncertainties, one needs to model these parameters as random variables. If the probability information (such as

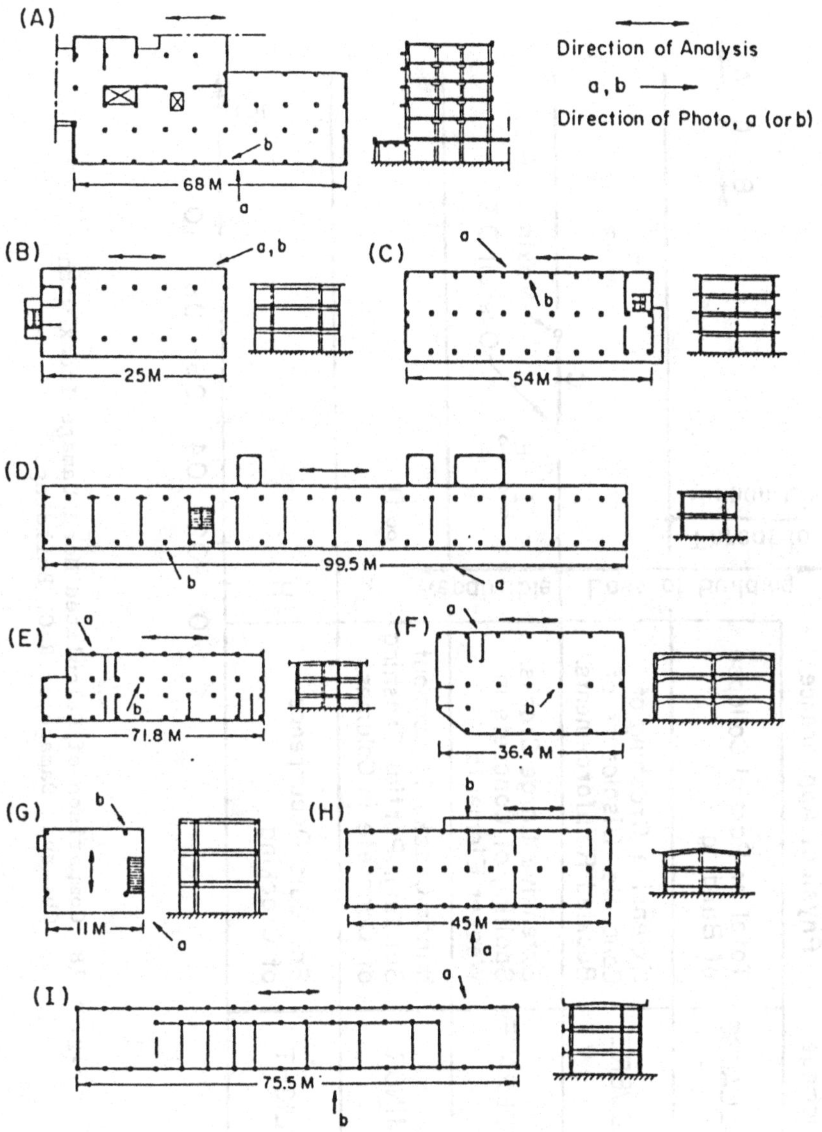

Fig. 5.17 Plan and Elevation of Buildings

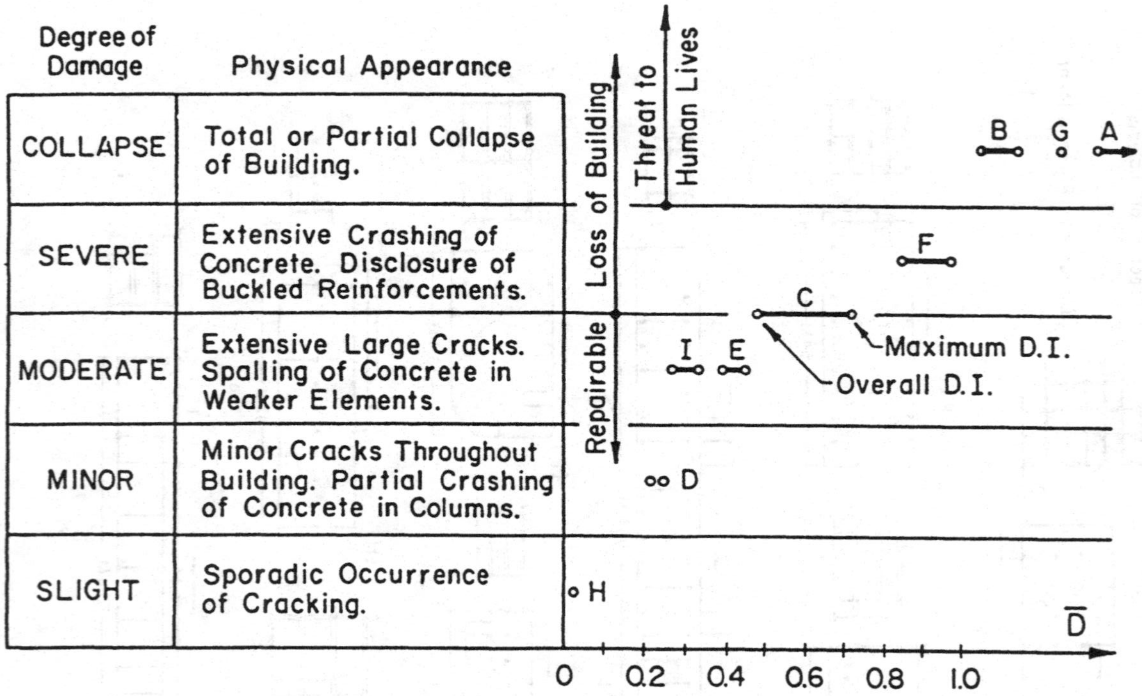

Fig. 5.18 Comparison of Calculated Mean Damage Index with Observed Damage of R.C. Buildings

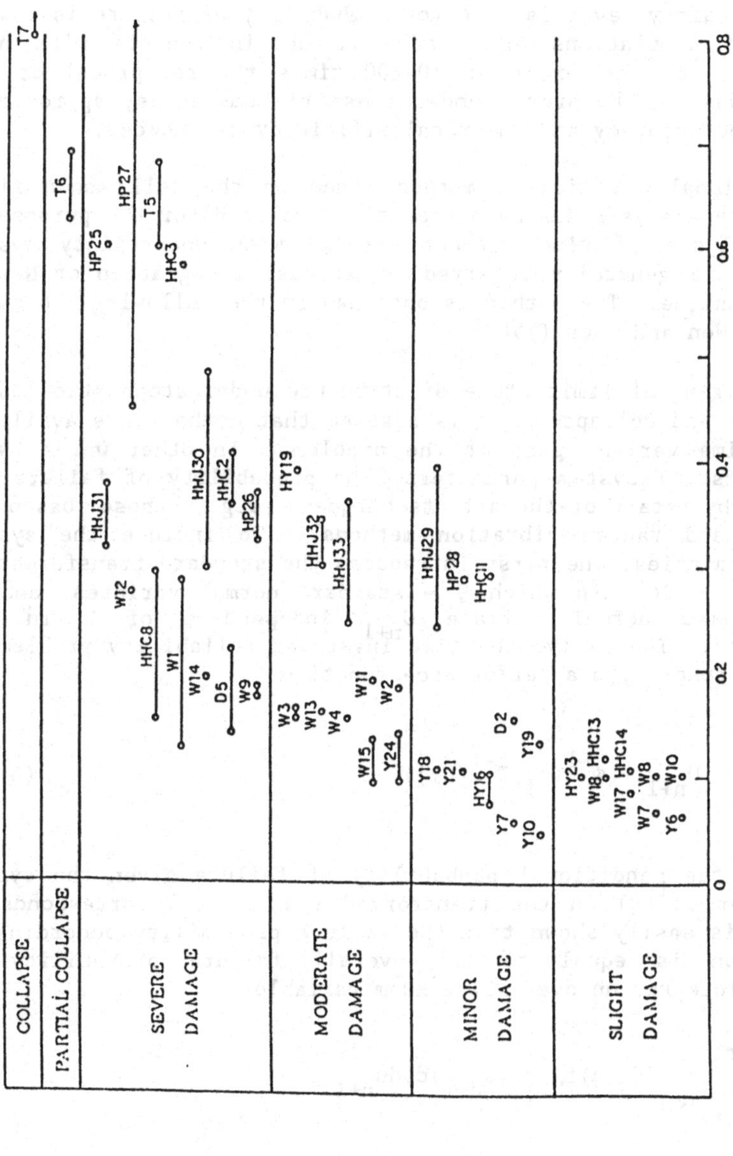

Fig. 5.19 Comparison of Calculated Mean Damage Index with Observed Damage of Masonry Buildings

density functions) of the parameters are available one can first solve the time variant part of the problem with given system parameters and then integrate over the system parameters to find the overall reliability. However, the computations could become excessive since repeated solutions of the structural response under stochastic loadings are required, in particular when the number of system parameters considered is large. Alternatively, one can use Monte-Carlo simulations. Again, when the safety level is high (or probability of failure is low) a large number of simulations are needed to obtain results with high confidence; e.g., of the order of 10-100 times the reciprocal of the failure probability. Therefore, under these circumstances, approximate methods with good accuracy and numerical efficiency are needed.

A computationally efficient method based on the well known first order reliability analysis has been recently developed for the purpose of reliability analysis of time variant systems with uncertainty system parameters. It is generally referred to as Fast Integration or Nested Reliability technique. The method is outlined in the following. Details can be found in Wen and Chen (1987).

Consider a general limit state of structure under stochastic loads, including damage and collapse. It is assumed that methods are available to solve the time variant part of the problem. In other words, with known (deterministic) system parameters, the probability of failure can be determined by state-of-the-art techniques, e.g., those based on random process and random vibration methods. To include the system parameter uncertainties, one first introduces the standard transformation of $\underset{\sim}{X}$ to $\underset{\sim}{U}$ by $\underset{\sim}{U} = T(\underset{\sim}{X})$ in which $\underset{\sim}{U}$ = standard normal variates, and an auxiliary standard normal variate U_{n+1} independent of $\underset{\sim}{U}$ in the transformed space. Then solve the time invariant reliability problem in the transformed space with a performance function

$$g_U(\underset{\sim}{U}, U_{n+1}) = U_{n+1} - \Phi^{-1}\{p_f[T^{-1}(U)]\} \tag{5.58}$$

in which p_f is the conditional probability of failure given the system parameters $\underset{\sim}{X}$, or $T^{-1}(\underset{\sim}{U})$ in the transformed space. $g<0$ corresponds to "failure". It is easily shown that the failure probability according to this formulation is equal to the overall failure probability by performing the integration over all random variables.

$$p_f = \int_{g(\underset{\sim}{u}, u_{n+1})<0} f_{\underset{\sim}{U}}(\underset{\sim}{u}) f_{U_{n+1}}(u_{n+1}) d\underset{\sim}{u} du_{n+1}$$

$$= \int_{\underset{\sim}{u}} f_{\underset{\sim}{U}}(\underset{\sim}{u}) \left\{ \int_{\infty}^{\Phi^{-1}(p_f)} f_{U_{n+1}}(u_{n+1}) du_{n+1} \right\} d\underset{\sim}{u}$$

$$= \int_{\underset{\sim}{u}} f_{\underset{\sim}{U}}(\underset{\sim}{u}) p_f[T \quad (\underset{\sim}{u})] d\underset{\sim}{u}$$

$$= \int_{\underset{\sim}{x}} f_{\underset{\sim}{X}}(\underset{\sim}{x}) p_f(\underset{\sim}{x}) d\underset{\sim}{x} \tag{5.59}$$

In searching for the design point in the first order method, the iterations can be easily done at the points along the limit surface since U_{n+1} is given explicitly in terms of $\underset{\sim}{U}$ in the limit state function (Eq. 5.58). The Jacobian matrix consists of only that associated with the system parameter $\underset{\sim}{X}$ transformation. The gradients of g_U at the point on the failure surface $g_U=0$ are

$$\left[\frac{\partial g_U}{\partial u_i}\right]_{g_U=0} = J_n^T[-\nabla p_f(\underset{\sim}{x})]/\phi(u_{n+1}) \qquad \text{for } i = 1,2,\ldots,n \tag{5.60}$$

$$\frac{\partial g_U}{\partial u_{n+1}} = 1 \tag{5.61}$$

in which J_n is the Jacobian matrix for transformation of $\underset{\sim}{X}$ to $\underset{\sim}{U}$, ϕ is the probability density function f the standard normal variable.

$$u_{n+1} = \Phi^{-1}[p_f(\underset{\sim}{x})] \tag{5.62}$$

To evaluate $\nabla p_f(\underset{\sim}{x})$, a finite difference scheme may be used if a closed form solution of $p_f(\underset{\sim}{x})$ is not available, which is likely o be the case of problems of some complexity. For example, if a central difference is used, to arrive at the design point and the solution one needs to solve for $p_f(\underset{\sim}{x})$ (the time variant part of the problem presumably most time-consuming), only 2nr times in which n is the dimension of $\underset{\sim}{X}$ and r is the number of iterations required. Compared to other methods such as numerical integration and Monte-Carlo simulation the computation required of this method is much less; it is, therefore, referred to as Fast Integration method. It is a method of conditioning of the system parameters and is sometimes referred to as a nested reliability technique. It can be applied to time variant problems where performance functions and response (random) variables as required in traditional reliability formulation are difficult to identify such as those involving damage and collapse of redundant systems. Details of application of this method to a variety of time variant systems can be found in Wen and Chen (1987).

REFERENCES

Atalik, T. S. and S. Utku. "Stochastic Linearization of Multi-Degree-of-Freedom Nonlinear Systems," Earthquake Engineering and Structural Dynamics, Vol. 4, 1976.

Baber, T. T. and Y. K. Wen. "Stochastic Equivalent Linearization for Hysteretic, Degrading Multistory Structures," Civil Engineering Studies, SRS No. 471, University of Illinois, Urbana, Illinois, December 1979.

Baber, T. T. and Y. K. Wen. "Random Vibration of Hysteretic Degrading Systems," Journal of Engineering Mechanics Division, ASCE, Vol. 107, No. EM6, December 1981, pp. 1069-1087.

Bartels, R. H. and G. W. Stewart. "Solution of the Matrix Equation AX + XB = C," Algorithm 432 in Communications of the ACM, Vol. 15, No. 9, September 1972.

Casciati, F. and L. Faravelli. "Stochastic Equivalent Linearization in 3-D Hysteretic Frames," Proceedings, 9th SMiRT, M20-4, 1987.

Caughey, T. K. "Random Excitation of a System with Bilinear Hysteresis," Journal of Applied Mechanics, Transactions of the ASME, Vol. 27, December 1960.

Caughey, T. K. "Equivalent Linearization Techniques," Journal of the Acoustical Society of America, Vol. 24, No. 11, 1963.

Chen, X-W. About Atalik and Utku's Linearization Technique, Innsbruck University, Innsbruck, Austria, 1987.

Faravelli, L.; F. Casciati; and M. P. Singh. "Stochastic Linearization Algorithms and Their Applicability to Hysteretic Systems," Meccanica, Journal of the Italian Association of Theoretical and Applied Mechanics, Vol. 23, 1988, pp. 107-112.

Grigoriu, M.; S. E. Ruiz; and E. Rosenblueth. "The Mexico Earthquake of September 19, 1985--Nonstationary Models of Seismic Ground Acceleration," Earthquake Spectra, Vol. 4, No. 3, pp. 551-568, 1988.

Iwan, W. D. "A Generalization of the Method of Equivalent Linearization," International Journal of Nonlinear Mechanics, Vol. 8, 1973.

Iwan, W. D. and L. D. Lutes. "Response of the Bilinear Hysteretic System to Stationary Random Excitation," Journal of the Acoustical Society of America, Vol. 43, No. 3, 1968.

Iwan, W. D.; M. A. Moser; and L. G. Pararizos. "The Stochastic Response of Strongly Nonlinear Systems with Coulomb Damping Elements," Proceedings IUTAM Symposium on Nonlinear Stochastic Dynamic Engineering Systems, Innsbruck, Austria, June 1987, pp. 455-466.

Kwok, Y. H.; A-H. S. Ang; and Y. K. Wen. "Seismic Damage Analysis and Design of Unreinforced Masonry Buildings," Civil Engineering Studies, SRS No. 536, University of Illinois, June 1987.

Lin, Y. K. "Probabilistic Theory of Structural Dynamics," McGraw-Hill, New York, 1967.

Mark, W. D. "Power Spectrum Representation for Nonstationary Random Vibration," Random Vibration -- Status and Recent Developments, (Ed., I Elishakoff and R. H. Lyon), Elsevier, pp. 211-240, 1986.

Park, Y. J.; A. H-S. Ang; and Y. K. Wen. "Seismic Damage Analysis of Reinforced Concrete Buildings," Journal of Structural Engineering, ASCE, Vol. III, No. 4, April 1985, pp. 740-757.

Park, Y. J.; Y. K. Wen; and A. H-S Ang. "Two-Dimensional Random Vibration of Hysteretic Structures," Journal of Earthquake Engineering and Structural Dynamics, Vol. 14, 1986, pp. 543-557.

Park, Y. J.; A. H-S. Ang; and Y. K. Wen. "Damage-Limiting Asiesmic Design of Buildings," Earthquake Spectra, Vol. 3, No. 1, February 1987, pp. 1-26.

Pires, J.E.A.; Y. K. Wen; and A. H-S. Ang. "Stochastic Analysis of Liquefaction Under Earthquake Loading," Civil Engineering Studies, SRS No. 504, University of Illinois at Urbana-Champaign, Urbana, Illinois, April 1983.

Powell, G. H. and P. F-S. Chen. "3-D Beam-Column Element with Generalized Plastic Hinges," Journal of Engineering Mechanics, ASCE, Vol. 112, No. 76, July 1986.

Spanos, P. D. "Stochastic Linearization in Structural Dynamics," Applied Mechanics Review, Vol. 34, No. 1, January 1981.

Sues, R. H.; Y. K. Wen; and A. H-S. Ang. "Stochastic Evaluation of Seismic Structural Performance," Journal of Structural Engineering Division, ASCE, Vol. 111, No. ST6, 1985, pp. 1204-1218.

Wen, Y. K. "Approximate Method for Nonlinear Random Vibration," Journal of Engineering Mechanics Division, ASCE, Vol. 101, No. EM4, August 1975.

Wen, Y. K. "Method for Random Vibration of Hysteretic Systems," Journal of Engineering Mechanics Division, ASCE, Vol. 102, No. EM2, April 1976.

Wen, Y. K. "Equivalent Linearization for Hysteretic Systems Under Random Excitations," Journal of Applied Mechanics, ASME, Vol. 47, No. 1, March 1980, pp. 150-154.

Wen, Y. K. "Method of Random Vibration for Inelastic Structures," Applied Mechanics Review, ASME, Vol. 42, No. 2, February 1989.

Wen, Y. K. and H-C. Chen. "On Fast Integration for Time Variant Structural Reliability," Probabilistic Engineering Mechanics, Vol. 2, No. 3, 1987, pp. 156-162.

Wen, Y. K. and C-H. Yeh. "Bi-Axial and Torsional Response of Inelastic Structures Under Random Excitation," Structural Safety, Vol. 6, 1989, pp. 137-152.

Yeh, C. H. and Y. K. Wen. "Modeling of Nonstationary Earthquake Ground Motion and Biaxial and Torsional Response of Inelastic Structures," Civil Engineering Studies, SRS No. 546, University of Illinois, August 1989.

Chapter 6

STATICS AND RELIABILITY OF MASONRY STRUCTURES

A. Baratta
University of Naples, Naples, Italy

Abstract

In this chapter, a mechanical model for masonry structures is introduced. Specifically, it is assumed that masonry can be idealized as a particular kind of "no-tension" material. The expected agreement of the model with the real behaviour of masonry buildings is discussed. In this approach, it is assumed that actual masonry may be endowed with a small degree of tensile strength, but that it exhibits brittle rupture under tension stresses. It is proved that the no-tension material model approximates actual stresses, in the sense of complementary energy.

6.1 INTRODUCTION

The problem of constructing the foundations of a structural theory which is suitable for masonry structures, and particularly for old - and/or ancient - buildings has recently been widely discussed. This is especially true in connection with the need to establish the most suitable criteria for formulating design techniques, in connection with the restoration and consolidation of the architectural heritage.

Following recent seismic events in Italy, and the need to repair the damage inflicted by earthquakes on existing buildings (most of which supported by a stone skeleton) it has been realized that available techniques for the analysis and consolidation of such structures are inadequate. Their development is not at a level comparable with the corresponding theoretical background that is available for modern structural typologies (concrete, steel and so on). Starting from this poor state-of-the-art, a concentrated research effort has begun, and has developed in recent years in Italy. The aim has been to establish basic principles, allowing a prediction of the quantitative behaviour of mechanical models. Eventually, in conjunction with personal experience, this work may help in understanding the behaviour of masonry buildings, in interpreting the symptoms of possible disease and in forecasting the effect of reinforcements.

The prevalent feature that characterizes such structures, and makes them dissimilar from actual concrete and steel structures, is quite definitely their inability to resist tensile stresses. The idea that structural theories for masonry should take into account this fact, as possibly the most important feature, was first introduced explicitly by Heyman in 1966 [1,2], who proved, after a number of practical studies carried on with special reference to monumental buildings, that proneness to disease or collapse is much more dependent on the activation of cracking mechanisms than on the probability of crushing in compression of masonry (the latter, in most cases, results from the masonry being poorly stressed in compression, prior to reching its ultimate resistance). On the other hand localized fractures do not usually affect the performance of the skeleton, as can be observed in most existing masonry buildings. In other words, fractures should be considered as a physiological aspect of the masonry material, unless they are so large as to compromise the local resistance of the material elements, or so well organized that a collapse mechanism may be activated.

The logical conclusion is that the material model should include fracturing as an intrinsic pattern for the stress-strain relationships. Moreover, the structural model should be sensitive to the presence of collapse mechanisms in the neighbourhood of the actual equilibrium configuration.

Some authors, including the present writer, [see e.g. 3-16], have tried to develop a formal theoretical framework for such phenomena, as they are commonly observed in most existing masonry structures. Instrumental control of actual buildings and laboratory tests have provided much experimental evidence which can be used to guide the theoretical work.

The basic assumption is that the material model, that is intended to be an "analogue" of real masonry, cannot resist tensile stress, but behaves elastically (indefinitely) under pure compression. It is noted that these conditions give a well defined specification of the admissible domain for stresses, but allow complete freedom for the path of fracture growth. In other words, in building up the stress-strain relationships for anelastic deformation (fractures, in this case), one is free to include the most appropriate assumptions. Therefore, one can speak of Standard

NRT (Not-Resisting-Tension) material, or of Nonstandard NRT material, according to the circumstance that the material is assumed to fracture according to a pattern similar to the Drucker's postulate, or not. In the NNRT (Non-Standard NRT) case, one can imagine even more patterns.

In establishing suitable fracture laws, one aims at achieving two main objectives: i)similarity of the model to the actual material; ii)simplicity of the formal theory, allowing mathematical models to be handled as easily as possible. In other words, one should tolerate some mismatch between the model behaviour and actual behaviour, so long as the loss in accuracy corresponds to a gain in the solution power and the overall correspondence between the theoretical model and the actual structure is preserved (eventually with the help of engineering judgement). A similar point of view, and a similar compromise, is encountered in the statical theory of reinforced concrete. Here a compromise between the "credibility" of the theory and the practical use of the model is, historically, fully established. It is possible that the NRT model is an efficient one for engineering decision purposes in the following sense. The assumed hypotheses lead to results that, although not reflecting completely the physical aspects of the real phenomena, can be interpreted in a unique way, enabling a prediction of the basic behaviour of masonry structures.

The largest source of uncertainty in the behaviour of masonry material arises from the very random character of the tensile strength. Moreover the_ cracking of masonry in tension is essentially a brittle rupture. As a result of these two features, tensile resistance cannot be relied upon in assessing the reliability of masonry buildings, and the NRT assumption is probably a necessary one. It is important, however, to obtain some insight into the relationships between NRT models and brittle-tension-resistant structures; the latter may, in principle, be a more "credible" model for masonry. It must be stressed that brittle constitutive laws yield structural models which are very hard to handle and -a very significant difficulty if one seeks safety criteria- do not enjoy a uniqueness in their solution.

6.2 THE ELASTIC-NO-TENSION MATERIAL

The following treatment is intended to present the most significant features of a theory aiming at a simple formulation of the behaviour of materials which have no capacity to resist tensile stresses (NRT material). The subject is treated with reference to plane stress, and is intended for applications to masonry plane walls.

6.2.1 <u>The domain of admissible stress states</u>

If compressive stresses are distinguished by negative stress-measures, assuming that $\boldsymbol{\sigma}$ is the stress tensor at a generic point P of the wall (Fig. 6.1) and denoting by σ_1 and σ_2 ($\sigma_2 \leq \sigma_1$) the

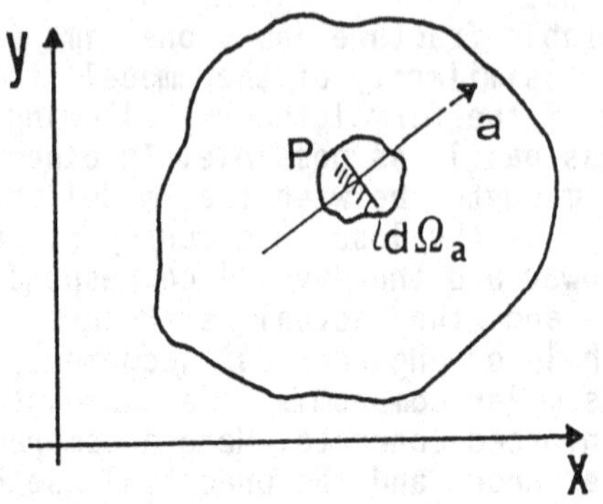

Fig. 6. 1: The material body under examination

respective principal stress components, admissible stress states should satisfy the condition

$$\sigma_1 \leq 0 \tag{6.1}$$

which, as well known, is equivalent to

$$\left.\begin{array}{l}\sigma_a \leq 0 \\ \\ \tau_a \text{ undefined}\end{array}\right] \quad a \in r_p \tag{6.2}$$

where the symbols σ_a and τ_a denote the orthogonal and tangential components, respectively, of the stress vector $\mathbf{t}_a$ on the plane element normal to the line "a" and r_p denotes the the set of lines passing through the point P.

Following these assumptions, a further hypothesis is made that the material behaves elastically in pure compression. The admissible domain for stresses, $\mathbf{K}_0$ say, in the plane of principal stress components, is therefore the negative quadrant (Fig. 6.2).

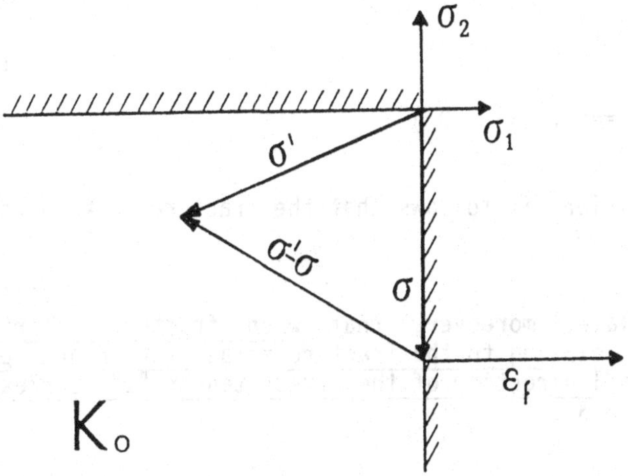

Fig. 6.2: The admissible stress domain for no-tension materials

6.2.2: Constitutive laws. stress-strain relationships

Since the material under examination is not able to resist tensile stresses, it is necessary to allow for the development of anelastic strain e_f, referred to in the following as the fracture strain. This strain has the role of transferring those forces deriving from unadmissible tensile stresses to the neighbouring material, in cases where the body has the capacity to achieve equilibrium with the same forces in pure compression.

Assuming that the displacements satisfy the conditions for the strain to be treated as infinitesimal, and denoting by **e** the total strain tensor at the point, it is possible to write

$$\mathbf{e} = \mathbf{e}_e + \mathbf{e}_f$$
$$\mathbf{e}_e = \mathbf{C}\sigma \tag{6.3}$$

where **C** denotes the usual tensor of elastic constants.

Note now that at any point where the fracture tensor is not zero, contact in the interior of the material is lost on a variety of plane elements. To take into account this phenomenon, let assume that the fracture strain is positive semidefinite and on any principal direction where the material is actually compressed, the corresponding coefficient of linear dilatation of the fracture strain must be zero. That is, in symbols

$$e_{fa} \geq 0 \quad \forall a \in r_p$$
$$(\sigma_i < 0) \iff (e_{fi} = 0) \tag{6.4}$$

From this condition, it follows that the fracture work is zero. In symbols

$$\boldsymbol{\sigma} \bullet \mathbf{e}_f = 0 \tag{6.5}$$

Let postulate, moreover, that <u>when fracture starts the displacements relating to the fracture strain tensor are parallel to the principal direction of the stress tensor "1" corresponding to $\sigma_1 = 0$ (Fig. 6.3).</u>

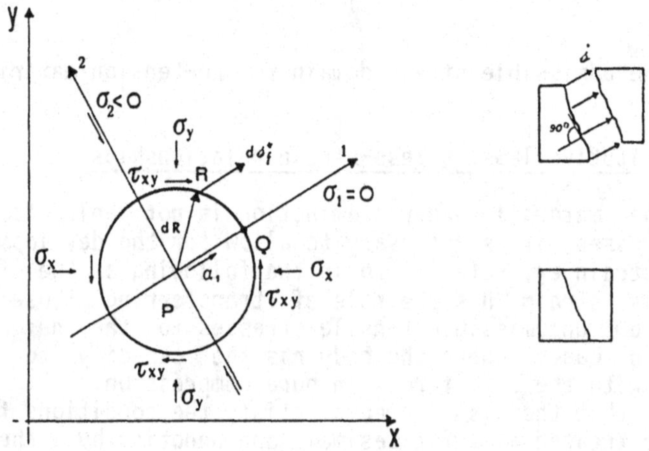

Fig. 6.3: Mechanism of activation of fractures

In symbols, let R be any point in the neighborhood of P, $\mathbf{ds}^*$ its displacement in pure fracture strain), $\mathbf{dR}$ its position with respect to P and $\boldsymbol{\alpha}_1$ the versor of the direction "1". One gets:

$$\mathbf{ds}^* = \mathbf{e}_f\mathbf{dR} = ds_R\,\boldsymbol{\alpha}_1 \tag{6.6}$$

with $ds_R > 0$.

From this postulate it follows that:

i) The principal directions of the fracture strain tensor are coincident with the directions of stress (that is $\mathbf{e_f}$ is coaxial with $\boldsymbol{\sigma}$).

In fact, for R ≡ Q, one gets

$$dQ = dr\, \boldsymbol{\alpha}_1$$

and from equation (6.6), with $\theta = ds_Q/dr$

$$\mathbf{e_f}\, \boldsymbol{\alpha}_1 = \theta\, \boldsymbol{\alpha}_1 \tag{6.7}$$

ii) The principal components of the fracture strain are nonnegative.

In fact, taking into account the statement i) and equation (6.4), one gets

$$\left.\begin{array}{l} (\sigma_2 < 0) \Longrightarrow (e_{f2} = 0) \\ (\sigma_1 = 0) \Longrightarrow (e_{f1} \geq 0) \end{array}\right] \tag{6.8}$$

iii) Since the principal strain components are the extremal values (maximum and minimum) of the dilatation in all possible directions, <u>the coefficient of linear dilatation of the fracture strain is nonnegative in any direction</u>

$$e_{fa} \geq 0 \quad \forall a \in r_p \tag{6.9}$$

iv) If $\boldsymbol{\sigma}$ is the effective stress tensor and $\boldsymbol{\sigma}'$ any other admissible stress state ($\boldsymbol{\sigma}' \in K_0$) one can infer that <u>the material obeys to the Drucker's stability postulate</u>

$$(\, \boldsymbol{\sigma}' - \boldsymbol{\sigma}\,) \bullet \mathbf{e_f} \leq 0 \tag{6.10}$$

In fact, recalling the statement ii) and equation (6.5), one can write

$$(\, \boldsymbol{\sigma}' - \boldsymbol{\sigma}\,) \bullet \mathbf{e_f} = \sigma'_1\, e_{f1} + \sigma'_2\, e_{f2} \leq 0$$

Summarizing, the fracture strain is characterized by nonnegative principal components and is coaxial with the stress state. If it is not zero, the principal directions of the fracture strain (say I and II) are, in contrast, quite free at the points of the body where stresses are zero.

In order to formulate explicitly the relationships that express the fracture strain as a function of the stress, let us consider initially the following proposition:

The space of symmetrical bidimensional second-order coaxial tensors is a two-dimensional vector space

Let, in fact, **A** and **B** be two symmetrical 2nd-order tensors, possessing common principal directions, whose versors are $\boldsymbol{\alpha}$ and $\boldsymbol{\beta}$. Let θ_1, θ_2 and μ_1, μ_2 be the eigenvalues of **A** and **B** respectively. Hence

$$\mathbf{A}\,\boldsymbol{\alpha} = \theta_1\boldsymbol{\alpha} \qquad \mathbf{B}\,\boldsymbol{\alpha} = \mu_1\boldsymbol{\alpha}$$

$$\mathbf{A}\,\boldsymbol{\beta} = \theta_2\boldsymbol{\beta} \qquad \mathbf{B}\,\boldsymbol{\beta} = \mu_2\boldsymbol{\beta} \tag{6.11}$$

Consider a third tensor **C** that has the same principal directions, with eigenvalues δ_1, δ_2

$$\mathbf{C}\,\boldsymbol{\alpha} = \delta_1\,\boldsymbol{\alpha} \qquad \mathbf{C}\,\boldsymbol{\beta} = \delta_2\boldsymbol{\beta} \tag{6.12}$$

Let k and h be two real numbers such that

$$\left.\begin{aligned} k\theta_1 + h\mu_1 &= \delta_1 \\ k\theta_2 + h\mu_2 &= \delta_2 \end{aligned}\right] \tag{6.13}$$

Such numbers exist and are uniquely defined provided that

$$\begin{vmatrix} \theta_1 & \mu_1 \\ \theta_2 & \mu_2 \end{vmatrix} \neq 0 \tag{6.14}$$

Consider now the tensor

$$\mathbf{D} = k\mathbf{A} + h\mathbf{B} \tag{6.15}$$

Hence

$$\mathbf{D}\boldsymbol{\alpha} = k\mathbf{A}\boldsymbol{\alpha} + h\mathbf{B}\boldsymbol{\alpha} = (k\theta_1 + h\mu_1)\,\boldsymbol{\alpha} = \delta_1\boldsymbol{\alpha} = \mathbf{C}\boldsymbol{\alpha}$$

(6.16)

$$\mathbf{D}\boldsymbol{\beta} = k\mathbf{A}\boldsymbol{\beta} + h\mathbf{B}\boldsymbol{\beta} = (k\theta_2 + h\mu_2)\,\boldsymbol{\beta} = \delta_2\boldsymbol{\beta} = \mathbf{C}\boldsymbol{\beta}$$

For every bidimensional vector $\mathbf{w}$, since $\boldsymbol{\alpha}$ and $\boldsymbol{\beta}$ are orthogonal, one gets

$$\mathbf{w} = w_\alpha\boldsymbol{\alpha} + w_\beta\boldsymbol{\beta} \tag{6.17}$$

whence

$$\mathbf{D}\mathbf{w} = w_\alpha\mathbf{D}\boldsymbol{\alpha} + w_\beta\mathbf{D}\boldsymbol{\beta} = w_\alpha\mathbf{C}\boldsymbol{\alpha} + w_\beta\mathbf{C}\boldsymbol{\beta} = \mathbf{C}\mathbf{w}; \quad \forall\, \mathbf{w} \in R^2 \tag{6.18}$$

and from equation (6.15) it follows that

$$\mathbf{C} = \mathbf{D} = k\mathbf{A} + h\mathbf{B} \tag{6.19}$$

This proves the thesis.

It is now possible to express the fracture strain through the superposition of two tensors, both coaxial to the stress tensor

$$\mathbf{e}_f = \theta_f \left[\boldsymbol{\sigma} - \frac{M_\sigma}{T_\sigma} \mathbf{I} \right] + \delta_f\, \mathbf{e}_f^* \tag{6.20}$$

where $\mathbf{I}$ is the unit tensor and $\mathbf{e}_f^*$ is the free fracture that can develop at the point when the stress state is zero. Moreover since $\boldsymbol{\sigma}\bullet\mathbf{e}_f = 0$

$$M_\sigma = \boldsymbol{\sigma} \bullet \boldsymbol{\sigma} \quad ; \quad T_\sigma = \sigma_1 + \sigma_2 \tag{6.21}$$

and θ_f, δ_f are parameters that control the activation of the different types of fracture strain, and are subject to the following conditions

$$\theta_f \sigma_1 = 0$$

$$M_\sigma \delta_f = 0$$

$$\theta_f \delta_f = 0 \qquad (6.22)$$

$$\theta_f \geq 0$$

$$\delta_f e^*_{f2} \geq 0$$

where it is intended that $\sigma_1 \geq \sigma_2$ and $e^*_{f1} \geq e^*_{f2}$

It is possible to verify that equation (6.20), with the conditions (6.22) satisfies all the properties i-iv stated in the above.

6.2.3 Structural analysis

The search for the stress field, in a panel of NRT material, under a system of active forces and settlements of the supports, can be pursued by employing the principle of the Minimum Complementary Energy.

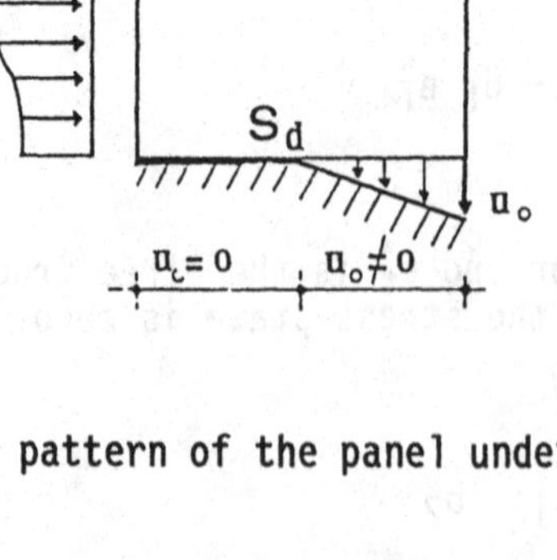

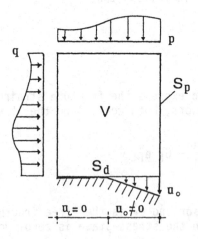

Fig. 6.4: The pattern of the panel under examination

Let

$$U(\sigma) = \frac{1}{2}\int_V \sigma \cdot C\sigma \, dV - \int_{S_d} T \cdot u_0 \, dS; \quad T = \sigma\, \alpha_n \quad \text{on } S_d \qquad (6.23)$$

be the functional of the Complementary Energy defined on the set D_0 of the admissible stress fields in equilibrium with the applied loads.

It can be proved that, if σ_0 is the solution stress field (i.e. the stress field such that the elastic strains $C\sigma_0$ can be made compatible with a continuous displacement field, by the superposition of a fracture strain field), the following condition holds

$$U(\sigma_0) = \min_{\sigma \in D_0} U(\sigma) = U_0 \qquad (6.24)$$

It can be proved that, because of the differentiability and the strict convexity of U, if D_0 is not empty, i.e. at least one statically admissible stress field exists, the solution σ_0 exists and it is unique.

Displacement and strain fields can be found, on the other hand, by the principle of the Minimum Total Potential Energy.

Let **B** be the differential operator that relates the displacement field to the strain tensor

$$\mathbf{e} = \mathbf{B}\,\mathbf{u} \qquad (6.25)$$

and

$$E(\mathbf{u}, \mathbf{e}_f) = \frac{1}{2}\int_V (\mathbf{B}\mathbf{u} - \mathbf{e}_f)\, \mathbf{D}\, (\mathbf{B}\mathbf{u} - \mathbf{e}_f)\, dV - \int_{S_p} \mathbf{p} \cdot \mathbf{u} \, dS \qquad (6.26)$$

be the functional of the Total Potential Energy of the wall. It can be proved [16,17] that the solution $\mathbf{u}_0$, $\mathbf{e}_{f0}$ satisfies following condition

$$E(\mathbf{u}_0, \mathbf{e}_{f0}) = \min_{\substack{\mathbf{u} \in V \\ \mathbf{e}_f \in W}} E(\mathbf{u}, \mathbf{e}_f) = E_0 \qquad (6.27)$$

where V denotes the set of continuous displacement fields, and W the set of admissible (i.e. $\mathbf{e}_f \geq 0$) fracture strain fields.
From the numerical point of view, both principles can be applied by making use of the methods of Operational Research.

It is interesting to note that the existence of the solution is strictly related to the load pattern (see e.g. [16]). This can be easily seen, at first glance, by recognising that no solution can exist for a cylindrical body in NRT material loaded by a tension axial force. On the contrary, it is easy to check that the solution coincides with De Saint Venant's treatment if the same force acts in compression.

Generally speaking, denoting by $\mathbf{p}$ the load pattern acting on the boundary of the wall, by V the set of possible displacements and by V_0 the set of displacement fields $\mathbf{u}_f$ compatible with fracture strains ("Mechanisms of collapse")

$$\mathbf{e}_f = \mathbf{B}\,\mathbf{u}_f \tag{6.28}$$

$$V_0 = \{\ \mathbf{u}_f \in V : \mathbf{B}\mathbf{u}_f \geq 0\ \} \tag{6.29}$$

a necessary condition for the existence of the solution is expressed by

$$\int_{Sp} \mathbf{p}\bullet\mathbf{u}_f\, dS \leq 0 \qquad \text{for any } \mathbf{u}_f \in V_0 \tag{6.30}$$

Remembering that the material under examination is unable to dissipate energy, the above variational inequality can be tested by checking that "if any <u>kinematically sufficient</u> mechanism $\mathbf{u}_f$ exists, i.e. such that

$$\int_V \mathbf{p}\bullet\mathbf{u}_f\, dS > 0 \tag{6.31}$$

no solution can exist for the equilibrium of the NRT solid". This can be looked at as a "kinematical theorem" of Limit Analysis for NRT bodies.

On the other hand, assume that under the load pattern $\mathbf{p}$ a admissible stress field $\boldsymbol{\sigma}$ exists (i.e. $\boldsymbol{\sigma} \leq 0$) in equilibrium with $\mathbf{p}$. By the Principle of Virtual Work, one gets

$$\int_{Sp} \mathbf{p}\bullet\mathbf{u}_f\, dS = \int_V \boldsymbol{\sigma}\bullet\mathbf{e}_f\, dV \leq 0 \qquad \forall \mathbf{u} \in V_0 \tag{6.32}$$

The above inequality can be tested by checking that "if under the assigned loads **p** any statically admissible stress field **σ** exists, i.e. such that it is in equilibrium with the applied loads and admissible, no kinematically sufficient mechanism exists and the structure does not collapse". This can be looked at as a Static Theorem of Limit Analysis for NRT bodies.

From the above considerations it follows that the study of the existence and uniqueness of the solution only requires a suitable kind of limit analysis for the structure. Uniqueness of the solution, that holds for the stress field but not for displacements and strains, is nevertheless a very significant feature and allows one to obtain reliable results by checking structural safety through a comparison of the calculated stresses with admissible (or limit) ones.

6.3 A MECHANICAL MODEL FOR ELASTIC-BRITTLE MATERIAL

In the present section a particular mechanical model is formulated which is intended to describe the behaviour of an elastic-brittle material (for brevity, ELBR). This material is endowed with an evolutionary strength domain in such a way that the loss in the material continuity, when any limit value of the tensile stress is attained, is taken into account.

Since such model is intended to approximate closely to the real behaviour of masonry, it will be assumed that the limit strength in tension is a small fraction of the compressive one. This justifies the further assumption that the response in compression is purely elastic, provided (obviously) that the maximum compression stress is controlled by a comparison with the admissible one. It is assumed, therefore, that the brittle phase is activated only if the maximum tension stress exceeds the respective limit value. After such an event, the fracture strain is suddenly activated and the tension stress falls to zero on the cracked plane element. The latter will necessarily coincide with one of the principal directions of stress, as they are directed immediately before cracking. It is assumed, moreover, that after cracking, the material cannot develop shear stresses along the fracture direction.

In order to express analytically such behaviour, the following assumptions are introduced.

6.3.1 Limit domain of stress

Let σ_0' be the tension strength of the material, σ_1 and σ_2 the principal stresses. The condition for the resistance of the material at time T is expressed in the following form:

$$\sigma_1 \le \sigma_0'; \quad \sigma_2 \le \sigma_0' \quad \text{if } \max_{t\in(0,T)} \sigma_1(t) < \sigma_0'$$

$$\sigma_1 \le 0 \; ; \quad \sigma_2 \le \sigma_0' \quad \text{if } \max_{t\in(0,T)} \sigma_1(t) = \sigma_0' \tag{6.33}$$

$$\sigma_1 \le 0 \; ; \quad \sigma_2 \le 0 \quad \text{if } \max_{t\in(0,T)} \sigma_1(t) = \max_{t\in(0,T)} \sigma_2(t) = \sigma_0'$$

If one wants to get a view of the admissible stress domain in the plane of principal stresses, the domain for the integer material is plotted in Fig. 6.5.a, while the two possible transient situations for unidirectional fracture are plotted as dashed lines in Fig. 6.5.b, (the stable form of the domain after cracking on two or more directions is also plotted by full line).

6.3.2 Constitutive laws

It is assumed that the development of the fracture strain obeys the following behaviour:
i) At first cracking, the tensor of fracture strain velocity is

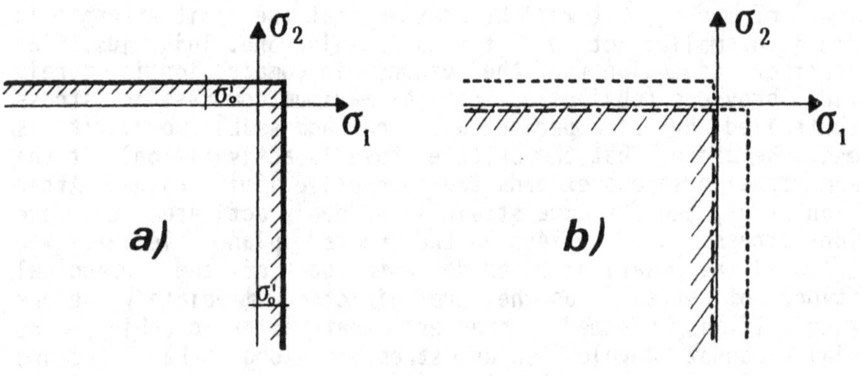

Fig. 6.5: The admissible stress domain for the ELBR material

coaxial with the stress state;
ii) The shear stress can act on a fracture direction only after the fracture strain has recovered to zero (i.e. to get a full elastic recovery linear and shear components of the fracture strain must be zero).

The two assumptions in above support the following postulates:

P1: $\sigma \bullet e_f = 0$ (6.34)

P2: $\sigma \bullet de_f - e_f \bullet d\sigma \geq 0$ (6.35)

In fact, in the evolution of the stress state and of cracking in the material 6 phases can be distinguished. These are analyzed briefly in the following:

F1) Integer material

In this phase the material behaves as a normal linearly elastic material. One has

$\sigma \neq 0; \quad d\sigma \neq 0$

$e_f = de_f = 0$

where the symbols $d\sigma$ and de_f denote variations in the stress state and in the fracture strain within a time interval (t,t+dt) for any variation of loads and/or of the strength capacity of the material, starting from the actual situation at time t, in which stresses and fractures are denoted by the symbols σ and e_f, respectively. Obviously both postulates P1 and P2 are satisfied.

F2) Material prone to fracture

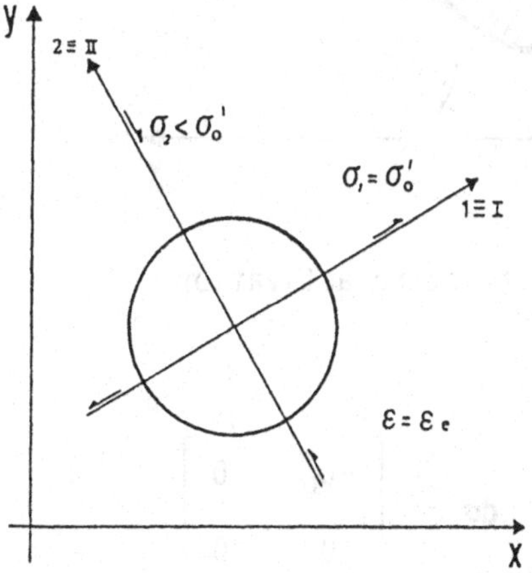

Fig. 6.6: The stress state before fracturing

This arises in a situation of unstable equilibrium. The tensile stress is at the limit value σ_0' on the principal direction "1" (Fig. 6.6) but the fracture strain is not yet active. The situation is essentially similar to F1.

F3) Development of the fracture

At this stage the fracture strain increases suddenly from a zero to a non-zero value, mainly due to the transformation in the fracture strain of the elastic dilatation on the principal stress direction "1". At the same time, the tensile stress σ_1 falls from the value σ_0' to zero. In agreement with assumption i), the fracture strain increment is coaxial both with respect to the instantaneous stress tensor and with its increment (Fig. 6.7).

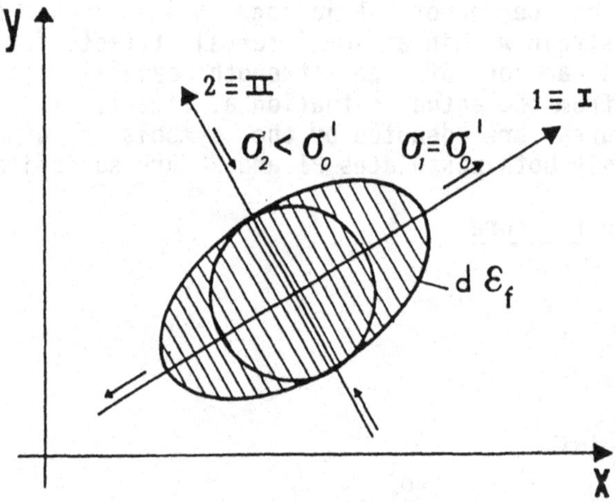

Fig. 6.7: First fracture activation

In symbols

$$\boldsymbol{\sigma} = \begin{bmatrix} \sigma_0' & 0 \\ 0 & \sigma_2 \end{bmatrix} \qquad \mathbf{d}\boldsymbol{\sigma} = \begin{bmatrix} -\sigma_0' & 0 \\ 0 & 0 \end{bmatrix}$$

$$\mathbf{e_f} = \begin{bmatrix} 0 & 0 \\ 0 & 0 \end{bmatrix} \qquad \mathbf{de_f} = \begin{bmatrix} \sigma_0'/E & 0 \\ 0 & 0 \end{bmatrix}$$

whence the relationships

$$\mathbf{e_f} \bullet \mathbf{d\sigma} = 0 \;; \quad \boldsymbol{\sigma} \bullet \mathbf{de_f} > 0 \quad ; \quad \boldsymbol{\sigma} \bullet \mathbf{e_f} = 0 \qquad (6.36)$$

This allows one to check that the basic postulates are also satisfied, in this phase.

F4) Propagation of the stress variation and of fracture strain

In this phase the principal stress directions remain stable in that, in the fractured direction, because of the loss in the contact between the opposite sides of the fracture, both normal and shear stress are zero. One must take into account, on the other hand, the possibility that the principal directions I and II of the fracture strain can rotate (Fig. 6.8), by virtue of the action that the material element under examination undergoes from the surrounding body. Moreover, a variation of the stress state that

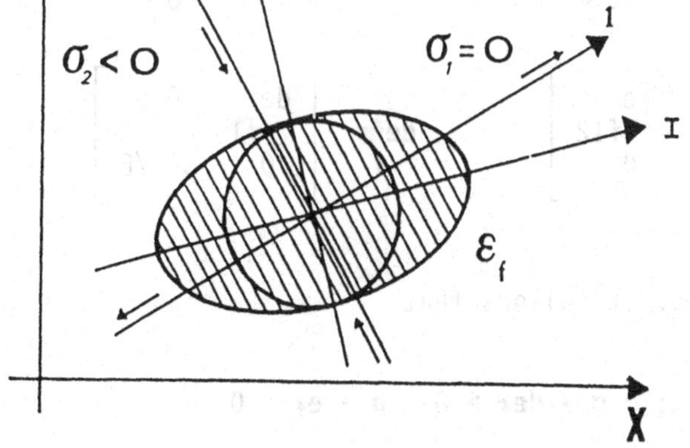

Fig. 6.8: The progress in fracture strain

involves only the principal component σ_2, in that σ_1 must remain zero, must be expected. Otherwise, the fracture is locked and the material behaviour falls to phase F6. Finally, one gets

$$\boldsymbol{\sigma} = \begin{bmatrix} 0 & 0 \\ 0 & \sigma_2 \end{bmatrix} \qquad \mathbf{d\sigma} = \begin{bmatrix} 0 & 0 \\ 0 & d\sigma_2 \end{bmatrix}$$

$$\mathbf{e}_f = \begin{bmatrix} e_{f1} & e_{f12} \\ e_{f12} & 0 \end{bmatrix} \qquad \mathbf{de}_f = \begin{bmatrix} de_{f1} & de_{f12} \\ de_{f12} & 0 \end{bmatrix}$$

with $e_{f1}>0$. It follows that

$$\mathbf{e_f} \bullet \mathbf{d\sigma} = 0 \quad ; \quad \boldsymbol{\sigma} \bullet \mathbf{de_f} = 0; \quad \boldsymbol{\sigma} \bullet \mathbf{e_f} = 0 \tag{6.37}$$

This is still in agreement with what was postulated.

F5) <u>Opening of a new fracture direction in the principal stress direction "2"</u>

In this case one gets

$$\boldsymbol{\sigma} = \begin{bmatrix} 0 & 0 \\ 0 & \sigma'_o \end{bmatrix} \qquad \mathbf{d\sigma} = \begin{bmatrix} 0 & 0 \\ 0 & -\sigma'_o \end{bmatrix}$$

$$\mathbf{e}_f = \begin{bmatrix} e_{f1} & e_{f12} \\ e_{f12} & 0 \end{bmatrix} \qquad \mathbf{de}_f = \begin{bmatrix} de_{f1} & 0 \\ 0 & \sigma'_o/E \end{bmatrix}$$

still with $e_{f1}>0$. It follows that

$$\mathbf{e_f} \bullet \mathbf{d\sigma} = 0; \quad \boldsymbol{\sigma} \bullet \mathbf{de_f} > 0; \quad \boldsymbol{\sigma} \bullet \mathbf{e_f} = 0 \tag{6.38}$$

in agreement with equations (6.35).

F6) Closure of fractures

Two cases can be distinguished:

a) When contact takes place again, the fracture strain is different from zero. In this case, throughout the fractured area the action of normal stress is allowed, but not of the shear stress, in line with assumption i). The principal stress directions remain coincident with the ones at the instant of the development of fractures. Assuming such directions as reference axes one gets

$$\boldsymbol{\sigma} = \begin{bmatrix} \sigma_1 & 0 \\ 0 & \sigma_2 \end{bmatrix} \qquad \mathbf{d\sigma} = \begin{bmatrix} d\sigma_1 & 0 \\ 0 & d\sigma_2 \end{bmatrix}$$

$$\mathbf{e}_f = \begin{bmatrix} 0 & e_{f12} \\ e_{f12} & 0 \end{bmatrix} \qquad \mathbf{de}_f = \begin{bmatrix} 0 & de_{f12} \\ de_{f12} & 0 \end{bmatrix}$$

with $\sigma_1 \leq 0$ and $\sigma_2 \leq \sigma_0'$. Hence the relationships:

$$\mathbf{e_f} \bullet \mathbf{d\sigma} = 0 \; ; \quad \boldsymbol{\sigma} \bullet \mathbf{de_f} = 0 \; ; \quad \boldsymbol{\sigma} \bullet \mathbf{e_f} = 0 \tag{6.39}$$

These are in agreement with the assumed postulates.

b) The fracture strain falls to zero. The material gains again the capacity to resist shear stresses in the fractured direction. The material fully recovers in the elastic phase and the fracture is locked. The principal stress direction can rotate,from now on.
Since in this phase $\mathbf{e_f} = \mathbf{de_f} = \mathbf{0}$, one gets, trivially,

$$\mathbf{e_f} \bullet \mathbf{d\sigma} = 0 \; ; \quad \boldsymbol{\sigma} \bullet \mathbf{de_f} = 0 \; ; \boldsymbol{\sigma} \bullet \mathbf{e_f} = 0 \tag{6.40}$$

3.3) Some Comments

The model discussed above should describe the behaviour of an ELBR material, so defined on the basis of the evolutionary properties of its strength domain.

In setting up the model, a number of hypotheses have been assumed, that one expects are suitable as an approximation to the behaviour of real materials. The main uncertainty is concerned with hypothesis ii); this was introduced as a material property, and seems rather doubtful. One should remember, however, that both hypotheses i) and ii) have been introduced in order to justify postulates P1 and P2, and that the latter are probably more realistic than the basic assumptions.

6.3.4 Structures in ELBR material

The existence of the solution for structures in an ELBR material has not been investigated so far. It is expected, anyway, that a solution exists at least for a class of load patterns, by analogy to what has been discussed for structures in NRT material. It can be shown, however, by some suitable examples, that the solution is not unique, unless the whole history of the loading process is fully specified.

In this section two properties of the solution are proven, that are useful in view of the discussion of the effectiveness of the NRT model for the analysis of masonry structures.

6.3.4.1 Monotonicity of the fracture process

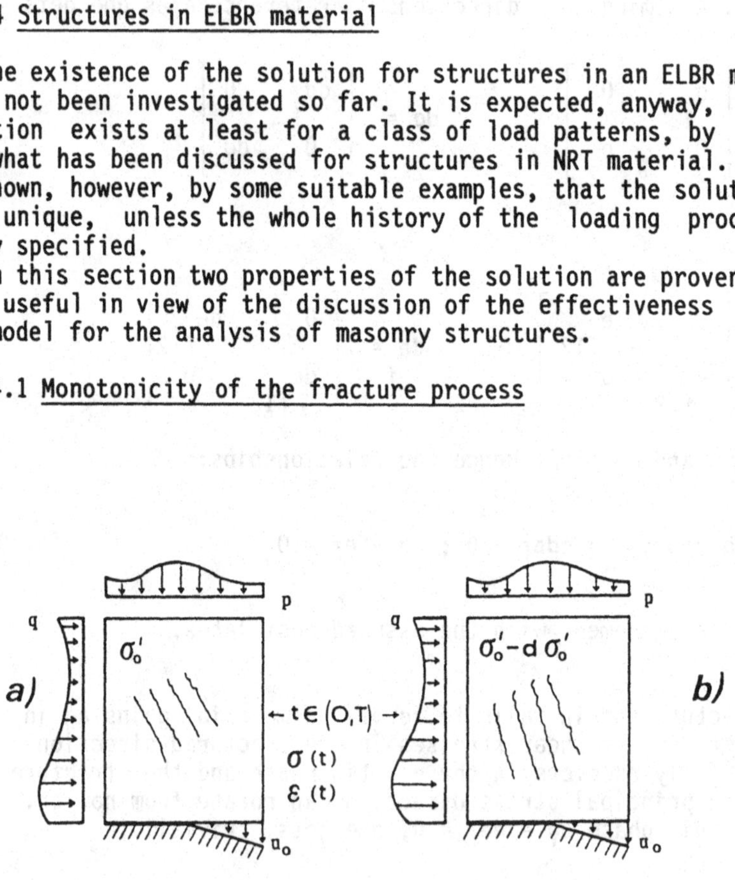

Fig. 6.9: The progress in fracturing by the decay in the tensile strength

Consider a panel made of ELBR material, in Fig. 6.9.a, and let σ'_0 be the limit strength in tension, $\mathbf{e}_f$ the fracture strain field, $\boldsymbol{\sigma}$ and $\mathbf{e}$ the stress and strain fields in solution. Let, moreover, $\boldsymbol{\sigma}_e$

be the stresses calculated under the hypothesis that the material behaves purely elastically.

Assume that, in the time interval (0,T), the limit tensile strength decays by the amount ?σ_0' Trivially one expects that such a loss produces a change in the stress/strain fields in the panel (Fig. 6.9.b) that can be viewed as a progressive process that evolves with time.

Let t be any instant in (0,T) and let $\boldsymbol{\sigma}(t)$, $\mathbf{e}(t)$, $\mathbf{e}_f(t)$ be the stress, strain and fracture fields at time t and $d\boldsymbol{\sigma}(t)$, $d\mathbf{e}(t)$, $d\mathbf{e}_f(t)$ be the respective variations in the next time increment dt. In any instant of the fracture process, the stress fields $\boldsymbol{\sigma}(t)-\boldsymbol{\sigma}_e(t)$ and $d\boldsymbol{\sigma}(t)$ are in self-equilibrium, whilst the strain increment $d\mathbf{e}(t)$ is compatible with null settlements of the supports. The Principle of Virtual Work yields

$$\int_V [\boldsymbol{\sigma}(t) - \boldsymbol{\sigma}_e] \bullet d\mathbf{e}(t)\, dV = 0 \tag{6.41}$$

with

$$d\mathbf{e}(t) = \mathbf{C}d\boldsymbol{\sigma}(t) + d\mathbf{e}_f(t) \tag{6.42}$$

whence, taking into account that

$$\mathbf{C}\,[\,\boldsymbol{\sigma}(t) - \boldsymbol{\sigma}_e] = \mathbf{e}(t) - \mathbf{e}_f(t) - \mathbf{e}_e \tag{6.43}$$

equation (6.41) becomes

$$\int_V \boldsymbol{\sigma}_e \bullet d\mathbf{e}_f(t)\, dV = \int_V [\,\boldsymbol{\sigma}(t) \bullet d\mathbf{e}_f(t) - \mathbf{e}_f(t) \bullet d\boldsymbol{\sigma}(t)]\, dV \tag{6.44}$$

By the postulate P2 one gets

$$\int_V \boldsymbol{\sigma}_e \bullet d\mathbf{e}_f(t)\, dV \geq 0 \tag{6.45}$$

Let $\delta\mathbf{e}_f$ be the variation of $\mathbf{e}_f$ produced by the loss in σ_0'. One gets

$$\int_V \boldsymbol{\sigma}_e \bullet \delta\mathbf{e}_f\, dV = \int_0^T \int_V \boldsymbol{\sigma}_e \bullet d\mathbf{e}_f(t)\, dV\, dt \geq 0 \tag{6.46}$$

By introducing the functional

$$F(e_f) = \int_V \sigma_e \bullet e_f \, dV \qquad (6.47)$$

and denoting by δF the variation of F after the loss in the tensile strength, one gets

$$\delta F = \delta \int_V \sigma_e \bullet e_f \, dV = \int_V \sigma_e \bullet \delta e_f \, dV \geq 0 \qquad (6.48)$$

Hence follows the first property: The functional $F(e_f)$ defined by equation (6.47) is a non-increasing function of the limit tensile strength.

6.3.4.2 Comparison property with the NRT solution

With reference to the mechanical pattern shown in Fig. 6.9, where V is the interior of the panel, S_p is the part of the boundary where surface tractions are specified, S_d is the part where settlements are prescribed, let

a) σ^*, e_f^* be the stress and fracture strains in the panel made of ELBR material;
b) σ_o, e_{of} the stress and fracture fields in the same plate made of NRT material;
c) σ_e the stress field in the plate made of linearly elastic material.

It is noted that the stress field $(\sigma_e - \sigma_o)$ is in self-equilibrium and that the strain field $(e^* - e_o)$ is compatible with zero settlements of the supports. The principle of Virtual Work yields

$$\int_V (\sigma_e - \sigma_o) \bullet (e^* - e_o) \, dV = 0 \qquad (6.49)$$

whence, taking into account that

$$e^* - e_o = C\,(\sigma^* - \sigma_o) + e^* - e_{of} \tag{6.50}$$

$$\sigma_o \bullet e_{of} = 0\ ; \quad C\sigma_o = e_o - e_{of}$$

and remembering that **C** is symmetrical, one obtains

$$\int_V (\sigma^* \bullet e_{of} - \sigma_o \bullet e^*_f)\, dV = \int_V \sigma_e \bullet (e_{of} - e^*_f)\, dV \tag{6.51}$$

Recalling equation (6.48) with $\delta e_f = e_{of} - e^*_f$ equation (6.51) yields

$$\int_V (\sigma^* \bullet e_{of} - \sigma_o \bullet e^*_f)\, dV \geq 0 \tag{6.52}$$

This expresses the fact that the work done by ELBR stresses, by the NRT fractures, is an upper bound to the work that NRT stresses would develop by ELBR fractures.

6.4 DISCUSSION OF THE NRT MODEL

Here the NRT model is used as an approximation to obtain some knowledge on the behaviour of structures made by materials that may depart (even significantly) from assumptions inherent in the model. The possibility of comparing the response of an NRT panel to the response of an ELBR one is investigated by assuming, conventionally, that the latter coincides with a real structure, whose behaviour is investigated by using the simpler NRT model. With reference to the scheme in Fig. 6.4, let consider the following four hypotheses for the material behaviour:

1) Indefinitely elastic material (EL)

2) Elastic-perfectly-plastic material with associated holonomous flow-law with an admissible domain as in Fig. 6.5.a (ELPL)

3) No-tension material (NRT)

4) Elastic-brittle material (ELBR)

In the cases 2) and 4) it is assumed that the limit strength in tension σ'_0 is only a small fraction of the maximum compression stress under normal loads.

In the following, the conditions that define the stress field in the solution, for the above four situations, are briefly summarized.

1) EL material

Let σ_e be the solution stress field under the hypothesis of purely elastic behaviour, **T** the reactions of the constraints on S_d, and D the set of all stress fields in equilibrium with the applied loads. The complementary energy functional is defined by

$$U = \frac{1}{2} \int_V \sigma \bullet C\sigma \, V - \int_{Sd} \mathbf{T} \bullet \mathbf{u_o} \, dS \tag{6.53}$$

and the solution σ_e must satisfy the condition

$$U(\sigma_e) = \min_{\sigma \in D} U(\sigma) = U_e \tag{6.54}$$

2) ELPL material

Let σ_p be the stress field in the solution, whose existence is assumed, and D_p the subset of D enjoying the property $\sigma_1 \le \sigma_o'$. One can write

$$U(\sigma_p) = \min_{\sigma \in Dp} U(\sigma) = U_p \tag{6.55}$$

Trivially, since $D_p \le D$, one gets

$$U_p \ge U_e \tag{6.56}$$

3) NRT material

Let σ_o be the solution stress field, whose existence is assumed, and D_o the subset of D_p enjoying the property $\sigma_1 \le 0$.
The principle of Minimum Complementary Energy yields

$$U(\sigma_o) = \min_{\sigma \in D_o} U(\sigma) = U_o \tag{6.57}$$

Since $D_o \leq D_p \leq D$,

$$U_o \geq U_p \geq U_e \tag{6.58}$$

4) ELBR material

Let $\boldsymbol{\sigma}^*$ be the solution stress field, and consider the relevant value of the complementary energy

$$U(\boldsymbol{\sigma}^*) = \frac{1}{2}\int_V \boldsymbol{\sigma}^* \bullet \mathbf{C}\boldsymbol{\sigma}^* \, dV - \int_{Sd} \mathbf{T}^* \bullet \mathbf{u}_o \, dS = U^* \tag{6.59}$$

In order to compare such a value with the one (U_o) corresponding to the NRT solution, consider the difference

$$DU = U^* - U_o = \frac{1}{2}\int_V (\boldsymbol{\sigma}^* \bullet \mathbf{C}\boldsymbol{\sigma}^* - \boldsymbol{\sigma}_o \bullet \mathbf{C}\boldsymbol{\sigma}_o)\, dV - \int_{Sd} (\mathbf{T}^* - \mathbf{T}_o) \bullet \mathbf{u}_o \, dS \tag{6.60}$$

By applying the Principle of Virtual Work to the stress field $\boldsymbol{\sigma}^* - \boldsymbol{\sigma}_o$, which is in self-equilibrium, and to the compatible strain field $\mathbf{e}^* = \mathbf{C}\boldsymbol{\sigma}^* + \mathbf{e}_f^*$ and remembering that $\boldsymbol{\sigma}^* \bullet \mathbf{e}_f^* = 0$, equation (6.60) can be written as

$$DU = \int_V \boldsymbol{\sigma}_o \bullet \mathbf{e}_f^* \, dV - \frac{1}{2}\int_V (\boldsymbol{\sigma}^* - \boldsymbol{\sigma}_o) \bullet \mathbf{C}(\sigma^* - \sigma_o)\, dV \tag{6.61}$$

Applying the P.V.W. again to the stress field $\boldsymbol{\sigma}^* - \boldsymbol{\sigma}_o$ and to the strain field $\mathbf{e}_o = \mathbf{C}\boldsymbol{\sigma}_o + \mathbf{e}_{of}$, and remembering that $\boldsymbol{\sigma}_o \bullet \mathbf{e}_{of} = 0$, the same equation (6.60) can be written

$$DU = \frac{1}{2}\int_V (\boldsymbol{\sigma}^* - \boldsymbol{\sigma}_o) \bullet \mathbf{C}(\boldsymbol{\sigma}^* - \boldsymbol{\sigma}_o)\, dV - \int_V \boldsymbol{\sigma}^* \bullet \mathbf{e}_{of} \, dV \tag{6.6}$$

Summing eqs. (6.61) and (6.62), one obtains

$$DU = \frac{1}{2} \int_V (\sigma_o \bullet e_f^* - \sigma^* \bullet e_{of})\, dV \qquad (6.63)$$

Hence, recalling equation (6.52)

$$DU \le 0 \;;\; U^* \le U_0 \qquad (6.64)$$

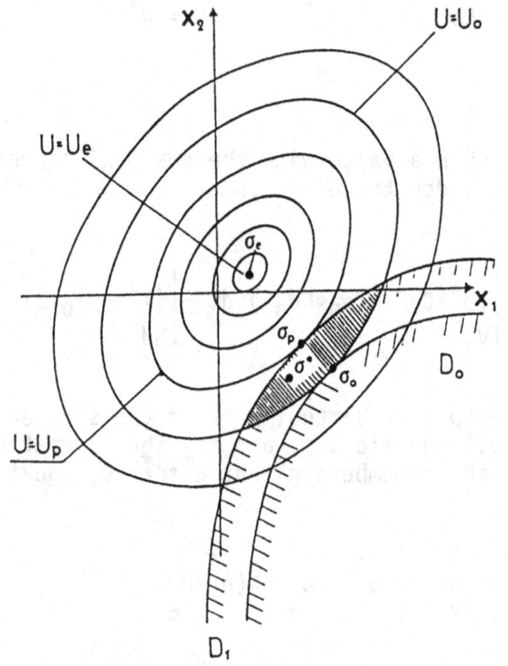

Fig. 6.10: The complementary energy scenario and possible solutions for different material behaviour

The complementary energy in the solution for an ELBR body is therefore bounded from above by the complementary energy that can be calculated in the solution for an NRT body, under the same actions.

It is obvious, from equation (6.55), that the same U^* is bounded from below by the energy U_p in the solution of the ELPL body.

If one wants to get a view of such a circumstance, assuming that the complementary energy depends only on two variables X_1 and X_2 (Fig. 6.10), one can conclude that the ELBR solution (conventionally assumed as the real material) must fall within the dashed area in the figure. This area, to some extent depending on the value of σ_0' can be considered as a neighborhood of σ_0. The NRT solution can then be viewed as an approximation to the actual stress field, in the sense that the distance between the two stress fields, expressed as the difference between the respective values of the complementary energy, is bounded by the difference of the energy corresponding to the ELPL solution - for the same limit strength in tension - and the one corresponding to the NRT solution; this difference tends to zero as σ_0' reduces to zero.

6.5 RELIABILITY OF MASONRY STRUCTURES

Let R(t) denote the reliability function of a building

$$R(t) = \text{Prob}\ \{\ \text{The structures survives up to time } t\ \} \tag{6.65}$$

with $\dot{R}(t) \leq 0$. One can interpret such a function as the degree of confidence that it is possible to associate with the expectation that the building shall not undergo significant malfunctions at that time. The hazard function may be defined by

$$h(t) = -\frac{\dot{R}(t)}{R(t)} \tag{6.66}$$

and can be interpreted as the conditional probability that the structure collapses at time t, given that it has survived so far. So, it can be interpreted as the degree of concern that one has to maintain when observing the behaviour of a building.

In the case of a masonry building, the most significant feature is the process of fracturing. The highest the velocity of opening of the fractures, the highest the degree of concern for the structure.

A measure of the velocity of opening of fractures is given by in the quantity

$$q(t) = \frac{1}{\int_V \sigma_e C \sigma_e \ dV} \int_V \sigma_e \cdot \dot{e}_f \ dV \tag{6.67}$$

If the problem is only concerned with the aging of the

structure, as is the case in the present paper, one can assume that the main factor is the decay in the tensile strength, with elapsed time. In this case, equation (5.3), by equation (6.67), from equation (6.45), is a semi-positive defined function of time, and may be considered an index of the hazard of the building.

A proposal to express the reliability function of the building can then be set up by

$$h(t) = q(t) = \frac{1}{\int_V \sigma_e C \sigma_e \, dV} \int_V \sigma_e \cdot \dot{e}_f \, dV \qquad (6.68)$$

Note that by equation (6.48), with $\delta e_f = e_{0f} - e_f$,

$$\int_V \sigma_e e_{0f} \, dV \geq \int_V \sigma_e e_f \, dV \qquad (6.69)$$

where e_{0f} are the NRT fractures.

Integration of equation (6.66) with

$$H(t) = \int_0^t h(t) \, dt = \frac{1}{\int_V \sigma_e C \sigma_e \, dV} \int_V \sigma_e \cdot e_f \, dV \qquad (6.70)$$

yields

$$R(t) = \exp\left[-H(t)\right] \qquad (6.71)$$

If the equilibrium condition (6.30) is verified for the NRT analogue structure, H(t) is bounded from above by

$$H_0 = \frac{1}{\int_V \sigma_e C \sigma_e \, dV} \int_V \sigma_e \cdot e_{0f} \, dV \qquad (6.72)$$

Hence it follows

$$R(t) \geq R_0 = \exp[-H_0] \tag{6.73}$$

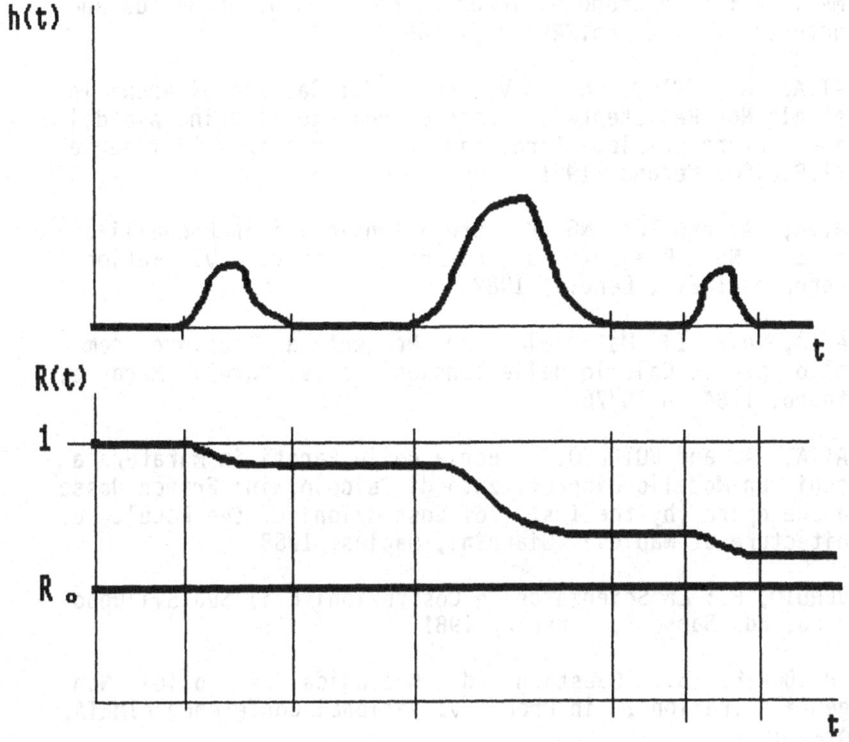

Fig. 6.11: Qualitative proceeding of reliability and hazard functions

Equation (6.73) means that the reliability function is endowed with a non-zero lower bound, if the NRT solution exists. Such a lower bound is higher, the smaller are the fractures introduced by the NRT assumption. On the contrary, when the NRT solution does not exist, or R_0 is so small that the lower bound may not exist, it may give a non-admissible measure of safety. In such cases, the growth of the hazard function should be seriously considered.

In Fig. 6.11 a possible evolution of both the hazard and the reliability function is sketched.

REFERENCES

1. HEYMAN, J.: The Safety of Masonry Arches, Int. Journ. of Mechanical Sciences, pp. 363-384, 1969

2. HEYMAN, J.: The Stone Skeleton, Int. Journ. of Solids and Structures, Vol. 2, pp.249-279, 1966

3. BARATTA, A., VIGO, M and VOIELLO, G.: Calcolo di Archi in Materiale Non Resistente a Trazione Mediante il Principio del Minimo Lavoro Complementare, in Proc. I National Conference ASS.I.R.C.CO. Verona, 1981

4. BARATTA, A. and TOSCANO, R.: Stati Tensionali in Pannelli di Materiale Non Reagente a Trazione, Proc. VI National Conference AIMETA, Genova, 1982

5. BARATTA, A.: Il Materiale Non Reagente a Trazione come Modello per il Calcolo delle Tensioni Nelle Pareti Murarie, Restauro, 1984, n.75/76

6. BARATTA, A. and VOIELLO,G: Teoria delle Pareti in Muratura a Blocchi: un Modello Discretizzato di Calcolo, in: Franco Jossa e la sua opera (by the Inst. of Costruzioni of the Faculty of Architecture of Naples), Giannini, Naples, 1988

7. BENVENUTO, E.: La Scienza delle Costruzioni e il Suo Sviluppo Storico, Ed. Sansoni, Firenze, 1981

8. DI PASQUALE, S.: Questioni di Meccanica dei Solidi Non Reagenti a Trazione, in Proc. VI National Conference AIMETA, Genova, 1982

9. DI PASQUALE, S.: Questioni Concernenti la Meccanica dei Mezzi Non Reagenti a Trazione, in Proc. VII Nat. Conf. AIMETA, Trieste, 1984

10. FRANCIOSI, V. and others: Il Procedimento delle Tensioni Ammissibili nella Verifica degli Archi e delle Volte in in Muratura in Zona Sismica Autostrade, n.9, 1982.

11. JOSSA, P. Sistemi Piani Semplici ed Armati di Materiale Non Reagente a Trazione, in Franco Jossa e la Sua Opera (by the Inst. of Costruzioni of the Faculty of Architecture of Naples), Giannini, Naples, 1988.

12. SACCHI LANDRIANI, G. and RICCIONI R. (Edrs): Comportamento Statico e Sismico delle Strutture Murarie, Ed. CLUP, Milano, 1982

13.SPARACIO, R.: Metodi di CAlcolo per le Costruzioni in Muratura, in: Il Consolidamento delle Costruzioni, CISM, Udine, 1982

14.VILLAGGIO, P.: Stress Diffusion in Masonry Walls, Journ. of Structural Mechanics, Vol.9, n.4, 1981

15.G. ROMANO and M. ROMANO: Elastostatics of Structures with Unilateral Conditions on Strains and Displacements, in: Unilateral Problems in Structural Analysis, Ravello (Italy), 1983

16.M. COMO and A. GRIMALDI: A Unilateral Model for the Limit Analysis of Masonry Walls, in: Unilateral Problems in Structural Analysis, Ravello (Italy), 1983

17.AUGUSTI, G., BARATTA, A. and CASCIATI, F.: Probabilistic Methods in Structural Analysis, Chapman & Hall, London, 1984

Chapter 7

ESSAY ON RELIABILITY INDEX, PROBABILISTIC INTERPRETATION OF SAFETY FACTOR, AND CONVEX MODELS OF UNCERTAINTY

I. Elishakoff
Florida Atlantic University, Boca Raton, FL, USA

Abstract

This paper represents a brief review of the concepts of the reliability of structures. Some closed form solutions, as well as approximate methods are elucidated. Different attempts to describe probabilistically the so-called "safety factor" are described and some instructive counter-examples are given. High sensitivity exhibited by the reliability of the structure is documented in a model structure. The same structure is also analyzed on the basis of non-probabilistic, convex modelling. The latter is complementary to the probabilistic methods when only limited information is available on uncertain variables and functions.

Introduction

The first harbinger of the new discipline of "probabilistic mechanics" appeared in Germany, in 1926, as the book by Mayer [1]. This method then was developed in Russia by Khozialov [2], Streletsky [3], Rzhanitsyn [4-5], in England by Tye [6], in France by Levi [7], in Sweden by Johnson [8], and fully flourished due to the efforts by Freudenthal, first in Israel and then in the United States [9,10] and many researchers around the world who followed these investigators. Now this subject has passed the age of adolescence [11] and became a widely accepted discipline with many monographs written and number of periodical journals appearing. In this paper, we will give a brief overview of the basics of probabilistic approach to structures. The pertinent but not exhaustive references are [12-37].

Basic Concepts

Consider the situation where the state of a structure in use can be described by a finite number of probabilistically dependent or independent parameters $X_1, X_2, \ldots, X_n$, part of which characterizes the loadings acting on the structure, and the other part is associated with the strength of the materials. For some combinations of its parameters the system is "acceptable" for use (in which case it is said to be in the safe state), whereas for other combinations it is "unacceptable" (in the failed state). The function $f(x_1, x_2, \ldots, x_n)$, x_j being the possible value of the random variable X_j may take on, which vanishes at the transition surface between the two states, is so defined that its positive values

$$f(x_1, x_2, \ldots, x_n) > 0 \tag{7.1}$$

represent the safe state, while its negative values

$$f(x_1, x_2, \ldots, x_n) < 0 \tag{7.2}$$

represent the failed state.

For example, if parameters $X_1, X_2, \ldots, X_m$ represent the strength of the materials, and $X_{m+1}, X_{m+2}, \ldots, X_n$ - the actual stresses, the failure surface could be put in the form

$$\begin{aligned} f &= R(x_1, x_2, \ldots, x_m) - S(x_{m+1}, x_{m+2}, \ldots, x_n) \\ &\equiv M(x_1, x_2, \ldots, x_n), \end{aligned} \tag{7.3}$$

with $M(x_1, x_2, \ldots, x_n)$ representing the "safety margin".

The reliability of the structure - the probability of its being in the safe state is obtained as

$$R = \int_0^{\infty} f_M(m)dm, \tag{7.4}$$

and the probability of failure, or the unreliability of the structure, as

$$P_f = \int_{-\infty}^{0} f_M(m)dm, \tag{7.5}$$

where $f_M(m)$ is the probability density of the safety margin which can be found through the familiar expression for the probability density of the difference of the random variables. Namely,

$$f_M(m) = \int_{-\infty}^{\infty} f_R(s + r) f_S(s) ds, \tag{7.6}$$

where $f_R(r)$ is the probability density of the strength and $f_S(s)$ - that of the stress. Irrespective of the specific densities of R and S, we have for the safety margin the mean

$$E(M) = E(R) - E(S). \tag{7.7}$$

For the uncorrelated R and S we get the following expression of the variance

$$Var(M) = Var(R) + Var(S). \tag{7.8}$$

The number of standard deviations of the safety margin in the interval S = 0 to s = E(S) is called reliability index (Fig. 1):

$$\beta = \frac{E(M)}{\sigma_M}, \tag{7.9}$$

where $\sigma_M = \sqrt{Var(M)}$ is the mean square deviation of the safety margin.

For the case where R and S are correlated, we have instead of Eq.7.9,

$$\beta = \frac{E(R) - E(S)}{[Var(R) - 2Cov(R,S) + Var(S)]^{1/2}} . \tag{7.10}$$

If R and S are normally distributed, then the reliability and probability of failure equal, respectively,

$$R = \Phi(\beta), \tag{7.11}$$

$$P_f = \Phi(-\beta), \tag{7.12}$$

where $\Phi(x)$ is the normal cumulative distribution function

$$\Phi(x) = \frac{1}{\sqrt{2\pi}} \int_{-\infty}^{x} e^{-t^2/2} dt. \tag{7.13}$$

Fig. 2 shows the iso-probability curves, ellipses for the general case of $\sigma_R \neq \sigma_S$. The reliability index β has an interesting geometrical interpretation using the standard independent normal variables, when the iso-probability curves become circles (Fig. 3):

$$R' = \frac{R - E(R)}{\sigma_R}, \quad S' = \frac{S - E(S)}{\sigma_S}. \tag{7.14}$$

3

The failure surface (3) is rewritten as

$$\sigma_R R' - \sigma_S S' + [E(R) - E(S)] = 0. \tag{7.15}$$

According to the analytical-geometry formula, the distance from the origin to the failure surface is

$$d = \frac{E(R) - E(S)}{\sqrt{\sigma_R^2 + \sigma_S^2}}, \tag{7.16}$$

which is formally identical with the expression for the reliability index β (Eq. 7.10) (Fig. 4). Point where d meets the failure surface is called the design point. Thus, we arrive at:

$$R = \Phi(\beta) = \Phi(d), \tag{7.17}$$

$$P_f = \Phi(-\beta) \doteq \Phi(-d). \tag{7.18}$$

There are few other situations in which exact expressions can be obtained for the reliability. Let, for example, the strength and stresses be independent, and the marginal densities exponential:

$$f_R(r) = \frac{1}{E(R)} \exp\left[-\frac{r}{E(R)}\right], \tag{7.19}$$

$$f_S(s) = \frac{1}{E(S)} \exp\left[-\frac{s}{E(S)}\right], \tag{7.20}$$

where E(R) and E(S) are the mean strength and stress, respectively. We obtain the following expression for the reliability:

$$R = \frac{E(R)}{E(R) + E(S)}. \tag{7.21}$$

Analogously, if R and S are independent random variables, having Rayleigh distribution

$$f_R(r) = \frac{\pi r}{2[E(R)]^2} \exp\left\{ -\frac{\pi r^2}{4[E(R)]^2} \right\}, \tag{7.22}$$

$$f_S(s) = \frac{\pi s}{2[E(S)]^2} \exp\{ - \frac{\pi s^2}{4[E(S)]^2} \} , \tag{7.23}$$

the reliability becomes

$$R = \frac{[E(R)]^2}{[E(R)^2 + [E(S)]^2} . \tag{7.24}$$

Additional important case is when both the stress and the strength have a log-normal distribution:

$$f_S(s) = \frac{1}{s\sigma_1\sqrt{2\pi}} \exp[- \frac{(\ell ns - a)^2}{2\sigma_1^2}], \ (s \geq 0) \tag{7.25}$$

$$f_R(r) = \frac{1}{r\sigma_2\sqrt{2\pi}} \exp[- \frac{(\ell ns - b)^2}{2\sigma_2^2}], \ (r \geq 0) \tag{7.26}$$

where a, b, σ_1 and σ_2 are the density parameters, so that

$$E(S) = \exp\ (a + \frac{1}{2}\sigma_1^2) ,$$

$$E(R) = \exp(b + \frac{1}{2}\sigma_2^2),$$

$$Var(S) = \exp(2a + \sigma_1^2)[\exp(\sigma_1^2)-1],$$

$$Var(R) = \exp(2b + \sigma_2^2)[\exp(\sigma_2^2)-1]. \tag{7.27}$$

The reliability is then

$$R = Prob(V = \frac{S}{R} \leq 1) = F_V(1), \tag{7.28}$$

where $F_V(v)$ is the probability distribution of the random variable V. Eq.7.28 may be rewritten as

$$R = \text{Prob}(\ell nV \le 0) = F_{\ell nV}(0). \tag{7.29}$$

Note that

$$\ell nV = \ell nS - \ell nR, \tag{7.30}$$

and since ℓnS and ℓnR both have a normal distribution, specifically ℓnS is $N(a_1\sigma_1^2)$ and ℓnY is $N(b_1\sigma_2^2)$, ℓnV is also normal, as a difference of normal variables $N(a-b,\ \sigma_1^2 + \sigma_2^2)$, implying that V is log-normal.

Reliability becomes

$$R = \Phi\left(-\frac{a-b}{\sqrt{\sigma_1^2+\sigma_2^2}}\right). \tag{7.31}$$

For other cases where exact solutions are obtainable, one should consult with the monographs by Ferry Borges and Castanheta [16], Rzhanitsyn [4], Bolotin [23], Ang and Tang [31], Augusti, Baratta and Casciati [32] and Elishakoff [29].

How does one calculate the reliability of a structure where exact solutions are unavailable? Such is usually the case if the relationship (1) is nonlinear. Under these circumstances, if the basic variables are still normally distributed, the formulas (7.17) and (7.18) are used, but now as approximations to the exact reliability and probability of failure, respectively (Fig.5). Equations (7.11) and (7.12), when the failure boundary is nonlinear, are commonly referred to as "Hasofer and Lind" index [19] on account of their systematic developments which lead to the wide-spread characterization of the reliability index as the minimal distance from the origin to the nonlinear failure surface. It appears instructive to give a quote from the paper by Shinozuka [30]: "... it is worth noting that the checking format for a modified design, recommended on the basis of these recent developments was in essence suggested by A. M. Freudenthal in this 1956 paper [10]. In this paper, referring to what is now known as the checking point, he wrote "because the critical condition (x^o,y^o) has the highest probability of occurance along the line r=0, it represents the combination to be used in design". The critical failure condition (x^o,y^o) indicates the point on the limit state (or failure) surface r=0 and located at the shortest distance, ρ, from the origin on the two-dimensional rectangular Carfesian coordinate space of the standardized Gaussian variables x and y. The critical point is also the point of maximum likelihood due to the Gaussian property assumed in the design variables. L. S. Lawrence [10] and J.M. Corso [10] in their discussion of the Frendenthal paper, pointed out, and Freudenthal concurred, that the limit state probability (probability of failure) is a function of the shortest distance, ρ. This is now known as the safety index, and can be obtained as $\Phi(-\rho)$, where $\Phi(\cdot)$ = the standardized Gaussian distribution function... his paper did suggest that the checking point used for design and that the checking point of shortest distance from the origin in a standardized Gaussian or transformed Gaussian variable space and at the same time is the point of maximum likelihood". Interestingly enough, the description of the minimum distance method appears in the monograph by Olszak et at [38]. Murzewski [37] is attributing this method to Levi [7]. In light of these comments, it appears more appropriate, to designate β in Eqs. 7.11 and 7.12 as minimum-distance reliability index.

How Accurate Is Minimum Distance Reliability Index?

To best answer this question, we will study the circular shaft-case amenable to exact solution [40]. Given a circular shaft of radius α subjected simultaneously to a bending moment M and torque T, characterized as random variables with joint probability density function $f_{MT}(m,t)$; (Fig. 6) the yield stress σ_y is constant with probability unity. According to the maximum shear stress theory of failure, the strength requirement reads

$$\frac{M_{eq}\, c}{I} \leq \sigma_y, \tag{7.32}$$

where M_{eq} is the "equivalent" moment:

$$M_{eq} = \sqrt{M^2 + T^2}, \tag{7.33}$$

$$R = \text{Prob}\left(\sqrt{M^2 + T^2} \leq \frac{\pi}{4}\sigma_y c^3\right). \tag{7.34}$$

Consider first the simplest case treated by Bolotin [18]: M and T are independent normal variables with zero means (a=b=0) and equal variances $\sigma_M = \sigma_T = \sigma$. Then the equivalent moment has a Rayleigh distribution

$$f_{M_{eq}}(m_{eq}) = \frac{m_{eq}}{\sigma^2}\exp\left(-\frac{m_{eq}^2}{2\sigma^2}\right), \tag{7.35}$$

with the attendant reliability

$$R = 1 - \exp\left(-\frac{\pi^2\sigma_y^2 c^6}{32\sigma^2}\right). \tag{7.36}$$

To compare this exact expression with the minimum distance method, we introduce the basic variables

$$Z_1 = M/\sigma, \quad Z_2 = T/\sigma. \tag{7.37}$$

The failure boundary becomes

$$Z_1^2 + Z_2^2 \le \rho^2, \tag{7.38}$$

where

$$\rho = \frac{\pi}{4}\frac{\sigma_y}{\sigma} c^3. \tag{7.39}$$

Eq.(7.38) represents a circle with radius ρ. The minimum distance to the circle equals the radius itself, so that $\beta = \rho$. Hence, under the minimum distance approximation we have

$$R = \Phi(\rho), \quad P_f = \Phi(-\rho), \tag{7.40}$$

whereas the exact solution (Eq. 7.36) in terms of ρ is

$$R = 1 - \exp(-\tfrac{1}{2}\rho^2), \quad P_f = \exp(-\tfrac{1}{2}\rho^2). \tag{7.41}$$

For highly reliable structures, which is where our interest lies, the approximate solution is remarkably close to the exact one.

Consider now the more realistic case $a^2 + b^2 \neq 0$ with the former restriction $\sigma_M = \sigma_T = \sigma$ still retained. In terms of the basic variables, the probabilistic counterpart of Eq.7.31 reads:

$$R = \text{Prob}[A \equiv (Z_1 + \frac{a}{\sigma})^2 + (Z_2 + \frac{b}{\sigma})^2 \le \rho^2]. \tag{7.42}$$

The random variable A has a non-central chi-square distribution with two degrees of freedom, and the noncentrality parameter

$$r = (a^2 + b^2)/\sigma^2. \tag{7.43}$$

Thus the reliability can be found via the extensive tables available. In the case $a/\sigma \gg 1$, $b/\sigma \gg 1$, an asymptotic expression is available, valid for $\rho > 5$:

$$R = \text{Prob}(A \le \rho^2) \simeq \Phi(\sqrt{\rho^2 - 1 - r}). \tag{7.44}$$

Let us compare this result to that obtained by the minimum distance method. The failure boundary is again a circle with radius ρ, but now centered at $(-a/\sigma, -b/\sigma)$. The minimum distance from the coordinate origin to the circle is (Fig.7):

$$d = \rho - r. \tag{7.45}$$

Hence

$$R \approx \Phi(\rho - r), \; P_f \approx \Phi[-(\rho - r)]. \tag{7.46}$$

Comparison of Equations (7.44) and (7.45) suggests that the minimum distance approximation is an excellent one, as for $\rho >> 1$ the asymptotically exact (Eq.7.44) and approximate expression (Eq.7.46) tend to each other.

Remarks on Safety Factor

Numerous attempts at probabilistic interpretation of the safety factor have been made in the literature despite the fact that the "spirits" of these two approaches are different. Before reviewing them, it is instructive to quote the following excerpts from popular textbooks concerning its definition:

a) "To allow for accidential overloading of the structure, as well as for possible inaccuracies in the construction and possible unknown variables in the analysis of the structure, a factor of safety is normally provided by choosing an allowable stress (or working stress) below the proportional limit".

b) "Although not commonly used, perhaps a better term for this ratio is factor of ignorance".

c) "The need for the safety margin is apparent for many reasons: stress itself is seldom uniform; materials lack the homogeneous properties theoretically assigned to them; abnormal loads might occur; manufacturing processes often impart dangerous stresses within the component. These and other factors make it necessary to select working stresses substantially below those known to cause failure".

d) "A factor of safety is used in the design of structures to allow for (1) uncertainties of loading, (2) the statistical variation of material strengths, (3) inaccurancies in geometry and theory, and (4) the grave consequencies of failure of some structures".

Freudenthal [20] remarks, "... it seems absurd to strive for more and more refinement of methods of stress-analysis if in order to determine the dimensions of the structural elements, its results are subsequently compared with so called working stress, derived in a rather crude manner by dividing the values of somewhat dubious material parameters obtained in conventional materials tests by still more dubious empirical numbers called safety factors". Indeed, it appears to the present author that in addition to its role as a "safety" parameter for the structure, it is intended as "personal insurance" factor of sorts for the design companies.

Probabilistic interpretation of the safety factor is not unique. We will discuss two of possible approaches towards such an interpretation. The "straightforward" safety factor itself, as the ratio R/S, is a random variable. The question is how to define it in probabilistic terms. A possible answer is the so-called central safety factor:

$$c = \frac{E(R)}{E(S)}, \tag{7.47}$$

which in certain situations is in direct correspondence with the reliability level. Indeed, for exponentially distributed strength and stress, the reliability, as per Eq.(21), reads

$$R = \frac{c}{1 + c}. \tag{7.48}$$

However, to achieve the reliability level of say, 0.999, the required central safety factor should be 999!

The situation is "better" for the case when R and S are independent Rayleigh distributed variables; in terms of the central safety factor c, Eq.(7.24) rewrites as

$$R = \frac{c^2}{1 + c^2}. \tag{7.49}$$

Under new circumstances, in order to achieve reliability of 0.999, the central safety factor should be $\sqrt{999} = 31.61$.

For normally distributed strength and stress, the safety index could be written as

$$\beta = d = \frac{c - 1}{\sqrt{\gamma_S^2 + c^2\gamma_R^2}}, \tag{7.50}$$

where $\gamma_S = \sigma_S/E(S)$, $\gamma_R = \sigma_R/E(R)$ are coefficients of variation of the stress and strength, respectively. The central safety factor corresponding to the reliability level r satisfies the quadratic

$$\omega_1 c^2 + \omega_2 c + \omega_3 = 0, \tag{7.51}$$

where

$$\omega_1 = 1 - \gamma_R^2 [\Phi^{-1}(r)]^2,$$

$$\omega_2 = -2,$$

$$\omega_3 = 1 - \gamma_S^2[\Phi^{-1}(r)]^2, \tag{7.52}$$

where $\Phi^{-1}(\cdot)$ is the inverse of $\Phi(\cdot)$. Under these circumstances, c depends on the reliability level r, but also on the coefficients of variation of the strength and stress.

Analogously, for the stress and strength, which have a log-normal distribution, the central safety factor is given by [29]:

$$c = \frac{\exp(a + \sigma_1^2/2)}{\exp(b + \sigma_2^2/2)}. \tag{7.53}$$

For $\sigma_1/a \ll 1$ and $\sigma_2 \ll 1$, Leporati [41] derived the following approximation

$$c = \exp\{\beta\, [(\frac{\sigma_1}{a})^2 + (\frac{\sigma_2}{b})^2]^{1/2}\}. \tag{7.54}$$

The alternative safety factor is introduced as follows

$$t = E(\frac{R}{S}) \tag{7.55}$$

instead of Eq.7.43. The following counter example is constructed by Elishakoff (see Ref.29, pp. 243-246), in which both R and S have an identical, uniform distribution over interval $[0,\alpha]$. Then on one hand, reliability is just one half, but the factor of safety t turns out to be infinity (!).

One can conclude, that for the reliability calculations one should have an information on the required reliability of the structure, with additional parameters E(R), E(S), σ_R, and σ_S specified, not necessarily in their direct connection with the "safety factor". In this connection it is instructive to quote Freudenthal [42]: "The predictive use, in structural design and analysis, of the theory of probability implies that the designer, on the basis of his professional competence, is able to draw valid conclusions from the probability figures obtained, so as to justify design decisions which in most cases, hinge on considerations of economy and utility. It is not implied that this use is in itself sufficient to make a design more reliable or more economical, any more than that the avoidance of the probabilistic approach makes it safer.

In fact, an approach based on the direct specification of a very low failure probability alone suffers from a major shortcoming: there is no intrinsic significance to a particular failure probability since no a priori rationalization can be given for the adoptation of a specific quantitive probability level in preference to any other, so that the selection of this level remains an arbitrary decision".

In the following section, we will illustrate the high sensitivity exhibited by the failure probability.

Sensitivity of Failure Probability

Consider an elastic bar compressed by an axial force. The uncertainty parameter is described by the nonvanishing eccentricities e_1 and e_2 of the force (Fig. 8).

The differential equation describing the deflection of the bar reads:

$$EI \frac{d^4w}{dx^4} + P \frac{d^2w}{dx^2} = 0 , \qquad (0 \leq x \leq L) \tag{7.56}$$

where EI is the flexural stiffness, P the axial force, w the displacement and L the length of the bar. Denoting

$$k^2 = \frac{P}{EI} . \tag{7.57}$$

The boundary conditions in terms of the bending moment are:

$$M_z (0) = Pe_1 \quad , \quad M_z (L) = Pe_2 . \tag{7.58}$$

Compliance with the boundary conditions yields the final expression for $M_z(x)$:

$$M_z(x) = - \frac{P}{\sin kL} (e_2 - e_1 \cos kL) \sin kx + P\, e_1 \cos kx. \tag{7.59}$$

The maximal bending moment M_z^* is

$$M_z^* (e_1, e_2) = \frac{P}{\sin kL} \sqrt{e_1^2 + e_2^2 - 2e_1e_2 \cos kL} . \tag{7.60}$$

This expression coincides with Eq. 1.44 in Pikovsky [43]. The problem of a bar in compression with two eccentricities was also studied by Young [44] and Timoshenko [45].

One can show that for the maximal bending moment to take place inside the bar, $0 < x^* < L$, the following conditions must be satisfied:

$$\cos kL < \frac{e_2}{e_1} < \frac{1}{\cos kL} , \tag{7.61}$$

and

$$0 < P < \frac{\pi^2 EI}{4L^2} , \quad 0 < kL < \frac{\pi}{2} \tag{7.62}$$

It can be shown that the maximal bending moment occurs inside the bar and condition (7.59) is dispensed with, in the following range of load variation:

$$\frac{\pi^2 EI}{4L^2} < P < \frac{\pi^2 EI}{L^2} \tag{7.63}$$

We assume now that the eccentricities constitute a random vector with a jointly exponential distribution and the following distribution function [46]:

$$F_{E_1E_2}(e_1, e_2) = \left[1 - \exp(-\frac{e_1}{\beta})\right] \left[1 - \exp(-\frac{e_2}{\gamma})\right] *$$

$$* \left[1 + \alpha \exp(-\frac{e_1}{\beta} - \frac{e_2}{\gamma})\right] , \tag{7.64}$$

where e_1 and e_2 take on only positive values and $\beta = E(E_1)$, $\gamma = E(E_2)$, $E(\ )$ denoting mathematical expectation.

For the sake of simplicity, we will concentrate on the case represented by Eq. 7.61. We are interested in the reliability of the bar, defined as the probability of nonexceedence of a limiting value m^* by the random variable M^*

$$R = \text{Prob}\left(M^* = \frac{P}{\sin kL} \sqrt{E_1^2 + E_2^2 - 2E_1E_2 \cos kL} \le m^*\right) . \tag{7.65}$$

The integration results in

$$F_{M^*}(m^*) = 1 - \frac{2\beta^3 + \alpha\beta^2\gamma - 5\beta^2\gamma - 2\alpha\beta\gamma^2 + 2\beta\gamma^2}{2\beta^3 - 7\beta^2\gamma + 7\beta\gamma^2 - 2\gamma^3} \exp(-\frac{m^* \sin kL}{\beta P})$$

$$- \frac{2\alpha\beta^2\gamma - 2\beta^2\gamma - \alpha\beta\gamma^2 + 5\beta\gamma^2 - 2\gamma^3}{2\beta^3 - 7\beta^2\gamma + 7\beta\gamma^2 - 2\gamma^3} \exp\left(- \frac{m^* \sin kL}{\gamma P}\right)$$

$$- \frac{\alpha\beta\gamma\,(2\gamma - \beta)}{2\beta^3 - 7\beta^2\gamma + 7\beta\gamma^2 - 2\gamma^3} \exp\left(- \frac{2m^* \sin kL}{\gamma P}\right)$$

$$- \frac{\alpha\beta\gamma\,(\gamma - 2\beta)}{2\beta^3 - 7\beta^2\gamma + 7\beta\gamma^2 - 2\gamma^3} \exp\left(- \frac{2m^* \sin kL}{\beta P}\right), \qquad (7.66)$$

when following restriction holds

$$2\beta^3 - 7\beta^2\gamma + 7\beta\gamma^2 - 2\gamma^3 \neq 0 .$$

In the particular cases where instead of the inequality in Eq.(7.65), we have an equality, the expressions for the reliability read:

$$F_{M^*}(m^*) = 1 - \left(1 + \frac{m^* \sin kL}{\beta P}\right) \exp\left(- \frac{m^* \sin kL}{\beta P}\right)$$

$$- \alpha \left[\left(\frac{m^* \sin kL}{\beta P} - 3 \right) \exp\left(- \frac{m^* \sin kL}{\beta P}\right) \right.$$

$$\left. + \left(\frac{2m^* \sin kL}{\beta P} + 3 \right) \exp\left(- \frac{2m^* \sin kL}{\beta P} \right) \right], \text{ for } \beta = \gamma. \qquad (7.67)$$

and

$$F_{M^*}(m^*) = 1 - 2\left(1 + \frac{\alpha}{3}\right) \exp\left(- \frac{m^* \sin kL}{\beta P}\right)$$

$$+ \left(1 + \frac{2\alpha m^* \sin kL}{\beta P}\right) \exp\left(- \frac{2m^* \sin kL}{\beta P}\right)$$

$$+ \frac{2\alpha}{3} \exp\left(- \frac{4m^* \sin kL}{\beta P}\right), \text{ for } \beta = 2\gamma . \qquad (7.68)$$

Finally

$$F_{M^*}(m^*) = 1 - 2(1 + \frac{\alpha}{3}) \exp(- \frac{m^* \sin kL}{2\beta P})$$

$$+(1 + \frac{\alpha m^* \sin kL}{\beta P}) \exp(- \frac{m^* \sin kL}{\beta P})$$

$$+ \frac{2\alpha}{3} \exp(- \frac{2m^* \sin kL}{\beta P}), \text{ for } \beta = \gamma/2 \quad . \quad (7.69)$$

It can be shown [46] that the following restrictions should be met, $F_M^*(m^*)$ to serve as the distribution function:

$$- 1 \le \alpha \le 1,$$

$$\alpha = 4\rho \ ,$$

where ρ is the coefficient of correlation between E_1 and E_2 .

Also, k on Eq. 2 could be expressed as

$$k = \frac{\pi}{L} \sqrt{\frac{P}{P_{cl}}} \ , \quad (7.70)$$

where P_{cl} is the classical buckling load of the simply supported bar $P_{cl} = \pi^2 EI/L^2$. The reliability equals

$$R = \text{Prob } (\Sigma \le \sigma_Y) = \text{Prob} \left[\frac{M^*c}{I} \le \sigma_Y\right]$$

$$= F_{M^*} \left[\frac{\sigma_Y I}{c}\right], \quad (7.71)$$

where Σ is the maximum stress, σ_Y-yield stress assumed to be constant, I-moment of inertia and c- the distance between the centroidal line and the extreme fiber where the maximum stress occurs. Reliability of the structure is obtainable from Eq. 7.66-7.69, by replacing m^* by $\sigma_Y I/c$.

Figures 9-11 depict the reliability of the structure versus the yield stress of the bar's material σ_Y. Fig. 9 is associated with P = 15 kN, P/P_{cl} = 0.569. The following data is used in Fig. 10: P = 20 kN, P/P_{cl} = 0.759, whereas in Fig. 11 the data is fixed at P = 23 kN, P/P_{cl} = 0.873. In all three figures β = 2mm, γ = 1.5 mm, and I/c = 1,333.3 mm^3; L = 1,000 mm, E = 200,000 MPa. As we see, the increase in the applied loading results in reduced reliability of the structure. The coefficient α is varied in Figures 9-11, in the range $-0.99 \leq \alpha \leq 0.99$. Whereas data on β and γ may be reliable, information on the coefficient α could be insuffificient, so that Figures 9-11 demonstrate the possible scatter in the reliability of the structure due to imprecise knowledge of parameter α. This is illustrated in Fig. 12, which addresses the design problem: Find the radius of the cross-section c so that the required codified reliability r, or codified probability of failure P_f^* = 1-r will be achieved.

If we fix value of P_f^* at 0.01, then, if the calculations are based on α = 0.99, the design value of the radius is c = 12.257; now, if the true value of α is -0.99, then the actual probability of failure of the system at this radius is $3.74 \cdot 10^{-3}$, i.e. is lower than the codified probability of failure. This implies that we had a case of the "favorable" imprecision. However, if our calculations are based on value α = -0.99, then the minimum required radius of the cross section should be c = 12.029. If however the true value of α is 0.99, then the chosen value of the radius corresponds to actual probability of failure 0.0234 instead of P_f^* = 0.01. This corresponds to the underestimation of the probability of failure by more than twice.

The situation is more severe for highly reliable structures. To get more insight we define the underestimation factor as the ratio of the actual-to-codified probabilities of failure

$$\eta = \frac{P_f}{P_f^*} \quad .$$

For $P_f^* = 10^{-3}$, the underestimation factor is over three; for $P_f^* = 10^{-4}$, η = 3.47; for $P_f^* = 10^{-5}$, η= 3.705 and finally for $P_f^* = 10^{-6}$ the underestimation factor reaches 3.82.

Thus, one would conclude that the system is acceptable for use, whereas the actual probability of failure is exceeding the codified one, and the system in fact is in a failed state, since the actual reliability is lower than the codified one, and the system in fact is in a failed state.

Under these circumstances of the high sensitivity of probability of failure, the natural question arises on how the probabilistic analysis could be used for design purposes. To attempt to answer this question, we will visualize that the cost C of production of the column is expressible as

$$C = \frac{q_1}{E(E_1) + E(E_2) + q_2}$$

where q_1 and q_2 are constants. Such a postulation maintains that more cost is associated with finer manufacturing, i.e. the one with less $E(E_i)$. Fig. 13 depicts the reliabilities of the columns associated with different mean imperfections $E(E_j)$, but their sum $E(E_1) + E(E_2)$ is kept constant. The Figure demonstrates that the maximum reliability is achieved for the equal mean imperfection parameters $E(E_1) = E(E_2)$. Thus, reliability studies could be utilized for comparative purposes;under other conditions being equal, one prefersthe manufacturing process which leads to higher reliability.

In the following section, we will devise an alternative non-probabilistic method to deal with uncertainty in the same problem.

Remarks on Convex Modelling of Uncertainty

Number of linear problems have been considered under set-theoretical, convex modelling of uncertainty in structures, put forward in monograph [47]. Particularly, impact failure of bars [48] and shells [49] was studied in detail, as was the response of a vehicle in uneven terrain [50]. By contrast, the only nonlinear problem considered in applied mechanics literature within the set-theoretical, convex modelling is buckling of shells with uncertain initial imperfections [51]. Ref.51 studied the first- and second-order approximation for the nonlinear function, since the exact solution was unavailable. Present paper contrasts the first- andsecond-order approximations discussed in detail in Ref. 46, with the exact analysis presented here for the first time, for the model structure of the bar with two eccentricities.

Consider again an elastic bar under an axial compressive force, applied with eccentricities e_1 and e_2. The maximum bending moment M_z^* is given by Eq. 7.60.

With e_1 and e_2 specified, the maximum value of the moment can be directly evaluated from Eq. 7.60. Assume now that the initial eccentricities are uncertain. In contrast to the previous section, we do not propose to model this uncertainty as randomness, under a probabilistic approach, but use an alternative, set-theoretical description, called in Ref. 46, "convexity modelling". The nominal values of the eccentricities are e_1^0 and e_2^0, respectively, and the deviations from these nominal values are denoted ζ_1 and ζ_2. We assume that these deviations vary within the ellipsoidal set:

$$Z(\alpha, \omega_1, \omega_2) = \{(\zeta_1,\zeta_2): (\frac{\zeta_1}{\omega_1})^2 + (\frac{\zeta_2}{\omega_2})^2 \le \alpha^2\}, \qquad (7.72)$$

where ω_1 and ω_2 are semi-axes of the ellipsoid, and α is its size parameter. We are interested in finding the maximum $\mu(\alpha,\omega_1,\omega_2)$, with respect to the uncertainty in the eccentricity, of the spacewise maximum bending moment

$$\mu(\alpha,\omega_1,\omega_2) = \max M_Z^* (e_1^0 + \zeta_1, e_2^0 + \zeta_2); \quad \{\zeta_1,\zeta_2 \in Z(\alpha,\omega_1,\omega_2)\}, \qquad (7.73)$$

where $\mu(\alpha,\omega_1,\omega_2)$ is the bending moment of the weakest bar in the ensemble Z. The maximum bending moment for uncertain eccentricities ζ_1 and ζ_2 to the first orderin ζ_1 and ζ_2 is

$$M_z^*(e_1^o + \zeta_1,\ e_2^o + \zeta_2) = M_Z^*(e_1^o,\ e_2^o) + \frac{\partial M_Z^*}{\partial e_1}\Big|_{e_i=e_i^o} \zeta_1 + \frac{\partial M_Z^*}{\partial e_2}\Big|_{e_i=e_i^o} \zeta_2. \tag{7.74}$$

We will evaluate the maximum bending moment as ζ_1 and ζ_2 vary in an ellipsoidal set $Z(\alpha,\omega_1,\omega_2)$. For convenience, we define the vector ϕ as follows:

$$\phi^T = \left\{ \frac{\partial M_Z^*}{\partial e_1}\Big|_{e_i=e_i^o},\ \frac{\partial M_Z^*}{\partial e_2}\Big|_{e_i=e_i^o} \right\}, \tag{7.75}$$

where the superscript T denotes matrix transposition. Eq. 7.73, in view of Eqs. 7.74 and 7.75 becomes

$$\mu(\alpha,\omega_1,\omega_2) = \max_{\zeta_1,\zeta_2 \in Z(\alpha,\omega_1,\omega_2)} [M_Z^*(e_1^o,e_2^o) + \phi^T\zeta], \tag{7.76}$$

where

$$\zeta^T = (\zeta_1,\ \zeta_2). \tag{7.77}$$

Define Ω as 2x2 diagonal matrix

$$\Omega = \begin{bmatrix} \frac{1}{\omega_1^2} & 0 \\ 0 & \frac{1}{\omega_2^2} \end{bmatrix}. \tag{7.78}$$

Then, Eq. 7.72 can be rewritten as

$$A(\alpha,\Omega) = \{\zeta:\ \zeta^T\Omega\zeta \le \alpha^2\}. \tag{7.79}$$

Eq. 7.76 calls for finding the maximum of the linear functional $\phi^T\zeta$ on the convex set $Z(\alpha,\omega_1,\omega_2)$. According to the well-known theorem (see e.g., Refs.52 and 53) a linear functional, considered on the convex set Z, assumes the maximum on the set of extreme points of Z. The latter is the collection of vectors $\sigma = (\zeta_1,\zeta_2)$ in the following set:

$$C(\alpha,\Omega) = \{\sigma: \quad \sigma^T\Omega\sigma = \alpha^2\}. \tag{7.80}$$

Thus the maximum bending moment becomes

$$\mu(\alpha,\Omega) = \max_{\sigma\in C(\alpha,\Omega)} [M_Z^*(e_1^0,e_2^0) + \phi^T\sigma]. \tag{7.81}$$

To solve the problem, we use the method of Lagrange multipliers. For details of derivation, the reader should consult with Ref. 54. Probabilistic analysis of the identical problem is given in Ref. 55.

For the maximum bending moment we arrive at the following expression

$$\mu(\alpha, \omega_1, \omega_2) = M_Z^*(e_1^0,e_2^0) + \phi^T\sigma_1 = M_Z^*(e_1^0, e_2^0) + \alpha\sqrt{\phi^T\Omega^{-1}\phi} \,. \tag{7.82}$$

In an analogous manner we arrive at the following expression for the minimum bending moment

$$\mu_{min}(\alpha,\omega_1,\omega_2) = M_Z^*(e_1^0, e_2^0) + \phi^T\sigma_2 = M_Z^*(e_1^0,e_2^0) - \alpha\sqrt{\phi^T\Omega^{-1}\phi} \,. \tag{7.83}$$

For the problem under consideration elements of vector ϕ can be found analytically:

$$\frac{\partial M_Z^*}{\partial e_1} = \frac{P\beta_1}{\sqrt{\beta_3}\ \sin kL}, \tag{7.84}$$

$$\frac{\partial M_Z^*}{\partial e_2} = \frac{P\beta_2}{\sqrt{\beta_3}\ \sin kL}, \tag{7.85}$$

where

$$\beta_1 = e_1 - e_2 \cos kL,$$

$$\beta_2 = e_2 - e_1 \cos kL,$$

$$\beta_3 = e_1^2 + e_2^2 - 2e_1e_2 \cos kL. \tag{7.86}$$

Hence the maximum bending moment reads

$$\mu_{max}(\alpha,\omega_1,\omega_2)=M_Z^*(e_1^0,e_2^0)+\alpha\sqrt{[\omega_1\frac{\partial M_Z(e_1,e_2)}{\partial e_1}\Big|_{e_i=e_i^0}]^2+[\omega_2\frac{\partial M_Z(e_1,e_2)}{\partial e_2}\Big|_{e_i=e_i^0}]^2}, \quad (7.87)$$

whereas the minimum bending moment is

$$\mu_{max}(\alpha,\omega_1,\omega_2)=M_Z^*(e_1^0,e_2^0)-\alpha\sqrt{(\omega_1\frac{\partial M_Z(e_1,e_2)}{\partial e_1}\Big|_{e_j=e_j^0})^2+(\omega_2\frac{\partial M_Z(e_1,e_2)}{\partial e_2}\Big|_{e_j=e_j^0})^2}. \quad (7.88)$$

The detailed second-order analysis can be found in Ref. 54.

We will show below that the maximum bending moment is a convex function of its arguments ζ_1 and ζ_2. Indeed, according to a theorem [52,53], a function of n variables defined on a convex set S is convex of and only if its Hessian matrix is positive semi-definite at all points in S. In our case the elements of the Hessian matrix γ_{ij} are

$$\gamma_{11} = (e_1^2 + e_2^2 - 2e_1e_2 \cos kL)^{-1/2}[1 - \frac{(e_1 - e_2 \cos kL)^2}{e_1^2 + e_2^2 - 2e_1e_2\cos kL}], \quad (7.89)$$

$$\gamma_{12} = - (e_1^2 + e_2^2 - 2e_1e_2\cos kL)^{-1/2}[\cos kL + \frac{(e_1 - e_2\cos kL)(e_2 - e_1\cos kL)}{e_1^2 + e_2^2 - 2e_1e_2\cos kL}], \quad (7.90)$$

$$\gamma_{22} = (e_1^2 + e_2^2 - 2e_1e_2\cos kL)^{-1/2} [1 - \frac{(e_2 - e_1 \cos kL)^2}{e_1^2 + e_2^2 - 2e_1e_2 \cos kL}]. \quad (7.91)$$

Direct calculation yields

$$\det[\Gamma] = \gamma_{11}\gamma_{22} - \gamma_{12}^2 \equiv 0. \quad (7.92)$$

Also,

$$\gamma_{11} > 0, \quad \gamma_{22} > 0. \tag{7.93}$$

Eqs.(7.92) and (7.93) imply that the function M_Z^* is convex. Therefore, we can apply the following theorem [52]: "Let f be a convex function defined on a bounded closed convex set S. If f has a maximum over S, this maximum is achieved at an extreme point of S."

Such being the case, we deduct that the maximum moment is achieved on the ellipse

$$\left(\frac{\zeta_1}{\omega_1}\right)^2 + \left(\frac{\zeta_2}{\omega_2}\right)^2 = \alpha^2 \tag{7.94}$$

We express ζ_2 from the latter equation as

$$\zeta_2 = \pm\,\omega_2 \sqrt{\alpha^2 - \frac{\zeta_1^2}{\omega_1^2}} \tag{7.95}$$

and substitute in Eq.(7.73), to yield

$$M_Z^*(e_1^0 + \zeta_1,\ e_2^0 + \zeta_2) = M_Z^*(e_1^0 + \zeta_1,\ e_2^0 \pm \omega_2 \sqrt{\alpha^2 - \zeta_1^2/\omega_1^2}\ ) \tag{7.96}$$

Now we seek the maximum of M_Z^* with respect to ζ_1 alone. For the maximum moment we demand that

$$\frac{\partial M_Z^*}{\partial \zeta_1} = 0. \tag{7.97}$$

This equation defines the ζ_1^* at which M_Z^* assumes maximum; then Eq.(7.95) determines the value of ζ_2^*. Substituting ζ_1^* and ζ_2^* into Eq.7.73, we obtain the maximum value of the bending moment.

Fig. 14 shows the variation of the bending moment over the region

$$\left(\frac{\zeta_1}{\omega_1}\right)^2 + \left(\frac{\zeta_2}{\omega_2}\right)^2 \le \alpha^2 \tag{7.98}$$

for $\omega_1 = 0.3$, $\omega_2 = 0.5$, at the nondimensional load level

$$\nu = \frac{P}{P_E}, \quad P_E = \frac{\pi^2 EI}{L^2} \tag{7.99}$$

equal $\nu = 0.3$. Fig. 15 is associated with $\omega_1 = 0.3$, $\omega_2 = 0.6$ and $\nu = 0.3$, whereas Fig. 16 illustrates the variation of M_Z^* for the values $\omega_1 = 0.3$, $\omega_2 = 0.6$ and $\nu = 0.5$. As we see in all these three instances the maximum value is achieved at the exterior point of the ellipse.

Figure 17 portrays the variation of the nondimensional maximal moment

$$\tilde{M}_Z^* = \frac{M_Z^*}{P_e c} \tag{7.100}$$

versus the ellipse's size α, where c is the radius of inertia of the bar's cross section. Moreover, $\omega_1 = 0.01$, $\omega_2 = 0.02$, $e^0 = 0.04$, $e^0 = 0.04$. The nondimensional load is fixed at $\nu = 0.3$. Broken curves 1 are associated with the first-order analysis, curves marked 2- with the second order analysis and curves marked 3 - with the exact results. The maximum moment increases with the size α of the uncertainty ellipse. For moderate values of α the agreement between the first-order, second-order and the exact analysis is excellent. It turns out that with the increase of the nondimensional applied load, the percentagewise disagreement between the first-order and the second-order analyses decreases considerably. On the other hand, for the larger nondimensional applied loads the difference between the low-order approximations and the exact analysis becomes wider.

We conclude, therefore, that the lower-order approximations yield acceptable results for small uncertainties and smaller loads. This modifies the conclusion, based on the comparison of the first-order and the second-order analyses, drawn in Ref. 46 that the first-order approximation was acceptable for small uncertainties and greater loads. It is remarkable that the similarity of the first-order and second-order approximations that occurs for the elastic bar does not generally suggest that these approximations are in good agreement with the exact solution.

Conclusion

This paper reviews some pertinent questions associated with uncertainty modelling in analysis of structures. It gives a critical appraisal of the probabilistic method and describes a new, non-probabilistic philosophy. The former is valid when plentiful information (probability densities) is available on the uncertain quantities involved. The latter, convex modelling is appropriate when the existing data is scarce and no valid probabilistic models can be constructed. Both of these approaches deal with different facets of uncertainty treatment. Theory of probability and random processes is not the only way to deal with uncertainty. Indeed, as Freudenthal [10] notes "Ignorance of the cause of variation does not make such variation random". These two methods appear to successfully complement each other, to make useful judgments based on reliably available experimental information.

Acknowledgement

This work was partially supported by the National Science Foundation through grant MSM-9015371. Any opinions, findings, and conclusions or recommendations expressed by this publication are those of the author and do not necessarily reflect the views of the National Science Foundation.

References

1. Mayer, H.: Die Sicherheit der Bauwerke and Ihre Berechnung nach Grenzkräften anstatt nach Zulässigen Spannungen, Springer, Berlin, 1926 (in German)(see also paper by Tichy, M., Max Mayer-Begründer der Berechnungsmethode nach Grenzzuständen, Technishe Mechanik, Magdeburg, Federal Republic of Germany, 1990, to appear).

2. Khozialov, N. F.: Safety Factors, Building Ind., Vol. 10, 1929, 840-844 (in Russian).

3. Streletskyi, N.S.: Statistical Basis of the Safety Factor of Structures, "Stroyizdat" Publishing House, Moscow, 1947 (in Russian).

4. Rzhanitsyn, A. R.: Design of Constructions with Materials' Plastic Properties Taken into Account, Gosudarstvennoe Izdatel'stvo Literatury Po Stroitel'stvu i Arkhitekture, Moscow, 1954, (second edition), Chapter 14 (in Russian). (see also a French translation: A R. Rjanitsyn: Calcul à la rupture et plasticite des constructions, Eyrolles, Paris, 1959).

5. Rzhanitsyn, A. R.: Theory of Reliability Design of Civil Engineering Structures, "Stroyizdat" Publishing House, Moscow, 1978 (in Russian).

6. Tye, W.: Factors of Safety- or of Habit?, Journal of Royal Aeronautical Society, Vol. 48, 1944, 487-494.

7. Levi, R.: Calculs probabilistes de la securite des constructions, Ann. Ponts et Chaussees, Vol. 119, No. 4, 1949, 493-539.

8. Johnson, A. I.: Strength, Safety and Economical Dimension of Structures, Bulletin No. 12, Royal Institute of Technology, Stockholm, 1953.

9. Freudenthal, A. M.: Safety of Structures, Transactions ASCE, Vol. 112, 1947, 125-180.

10. Freudenthal, A. M.: Safety and Probability of Structural Failure, Transactions ASCE, Vol. 121, 1956, 1337-1375.

11. Cornell, C. A.: Structural Safety: Some Historical Evidence That It is a Healthy Adolescent, in T. Moan and M. Shinozuka, "Structural Safety and Reliability", Elsevier Scientific Publishing Company, Amsterdam, 1981, 19-29.

12. Ekimov, V.V.: Probabilistic Methods in the Structural Mechanics of Ships, "Sudostroenie" Publishing House, Leningrad, 1966 (in Russian).

13. Cornell, C. A.: Probability-based Structural Code, ACI Journal, Vol. 66, 1969, 974-985.

14. Benjamin, J. R. and Cornell, C. A.: Probability, Statistics and Decision for Civil Engineers, McGraw Hill, New York, 1970.

15. Murzewski, J.: Bezpieczenstwo Konstrukeji Budowlanych, "Arkady" Publishing House, Warsaw, 1970 (in Polish).

16. Ferry Borges, J. and Castanheta, M.: Structural Safety, 2nd ed., National Civil Eng. Lab., Lisbon, Portugal, 1971.

17. Tichy, M. and Vorlicek, M.: Statistical Theory of Concrete Structures, "Academia" Publishing House, Prague, 1972.

18. Bolotin, V. V.: Application of the Methods of the Theory of Probability and the Theory of Reliability to Analysis of Structures, State Publishing House for Buildings, Moscow, 1971, (in Russian). English translation: FTD-MT-24-771-73, Foreign Technology Div., Wright Patterson AFB, Ohio, 1974.

19. Hasofer, A. M. and Lind, N. C.: Exact and Invariant Second-Moment Code Format, Journal of the Engineering Mechanics Division, Vol. 100, No. EM1, 1974, 111-121.

20. Ghiocel, D. and Lungu, D.: Wind, Snow and Temperature Effects on Structures Based on Probability, Abacus Press, Turnbridge Wells, Kent, UK, 1975.

21. Kapur, K. S. and Lamberson, L. R.: Reliability in Engineering Design, Wiley, New York, 1977.

22. Haugen, E. B.: Probabilistic Mechanical Design, Wiley-Interscience, New York, 1980.

23. Bolotin, V. V.: Wahrscheinlichkeitsmethoden zur Berechnung von Konstruktionen, VEB Verlag für Bauwesen, Berlin, 1981 (in German).

24. Ditlevsen, O.,: Uncertainty Modelling with Applications to Multidimensional Civil Engineering Systems, McGraw-Hill, New York, 1981.

25. Schuëller, G. I.: Einführung in die Sicherheit und Zuverlässigkeit von Tragwerken, Verlag von Wilhelm Ernst & Sohn, Berlin, 1981 (in German).

26. Thoft-Christensen, P. and Baker, M. J.: Structural Reliability Theory and Its Applications, Springer Verlag, Berlin, 1982.

27. Timashev, S.A.: Reliability of Large Mechanical Systems, "Nauka" Publishing House, Moscow, 1982 (in Russian).

28. Hart, G. C.: Uncertainty Analysis, Loads, and Safety in Structural Engineering, Prentice Hall, Inc., Englewood Cliffs, N.J., 1982.

29. Elishakoff, I.: Probabilistic Methods in the Theory of Structures, Wiley-Interscience, New York, 1983.

30. Shinozuka, M.: Basic Analysis of Structural Safety, ASCE Journal of Structural Engineering, Vol. 59, No. 3, 1983, 721-740.

31. Ang, A. H.-S. and Tang, W. H.: Probability Concepts in Engineering, Planning and Design, Vol. 2, Wiley, New York, 1984.

32. Augusti, G., Baratta, A. and Casciati, F.: Probabilistic Methods in Structural Engineering, Chapman and Hall, London, 1984.

33. Madsen, H. O., Krenk, S. and Lind, N. C.: Methods of Structural Safety, Prentice Hall, Englewood Cliffs, 1986.

34. Smith, G.N.: Probability and Statistics in Civil Engineering, Collins Professional and Technical Books, London, 1986.

35. Melchers, R. E.: Structural Reliability and Predictions, Ellis Horwood, London, 1987.

36. Ditlevsen, O.: Structural Reliability Methods, SBI, 1990 (in Danish).

37. Murzewski, J.: Niezawodnosc Konstrukcji Inzynierskich, Arkady, Warszawa, 1989 (in Polish).

38. Olszak, W., Kaufman, S., Eimer C. and Bychawski, Z.: Teoria Konstrukcji Sprezonych, Warszawa, Panstwowe Wydawnictwo Naukowe, 1961 (in Polish).

39. Elishakoff, I. and Hasofer, A. M.: On the Accuracy of Hasofer-Lind Reliability Index, Proceedings of ICOSSAR-85, International Conference on Structural Safety and Reliability, Kobe, Japan, 1985, Vol. 1, 229-239.

40. Elishakoff, I. and Hasofer, A. M.: Exact versus Approximate Analysis of Structural Reliability, International Journal of Thermal, Mechanical and Electromagnetic Phenomena in Continua, Vol. 44, 1987, 303-312.

41. Leporati, E.: The Assessment of Structural Safety, Research Studies Press, Forest Groves, Oregon, 1977.

42. Freudenthal, A., Introductory Remarks, in "International Conference on Structural Safety and Reliability", A. Freudenthal, ed., Pergamon Press, Oxford, 1972, pp. 5-6.

43. Pikovsky, A. A.: Statics of Column Systems with Compressed Elements, Gosudarstvennoe Izdatel'stvo Fiziko-Matematicheskoy Literatury, Moscow, 1961 (in Russian).

44. Young, D.M.: Stresses in Eccentrically Loaded Steel Columns, Publication of the International Association of Bridge Structural Engineers, Vol. 1, 1932.

45. Timoshenko, S. P. and Gere, J.M.: Theory of Elastic Stability, McGraw Hill, Auckland, 1963.

46. Gumbel, E. J.: Bivariate Exponential Distributions, American Statistical Association Journal, 1960, pp.698-707.

47. Ben-Haim, Y. and Elishakoff, I.: Convex Models of Uncertainity in Applied Mechanics, Elseiver Science Publishers, Amsterdam, 1990, pp. 1-43.

48. Ben-Haim, Y. and Elishakoff, I.: Dynamics and Failure of a Thin Bar with Unknown but Bounded Imperfections, in "Recent Advances Impact Dynamics of Engineering Structures" (D. Hui and N. Jones, eds)., AMD- Vol. 105, ASME, New York, 1989, 89-96.

49. Elishakoff, I. and Ben-Haim, Y.: Dynamics of a Thin Cylindrical Shell under Impact with Limited Deterministic Information on Its Initial Imperfections, Journal of Structural Safety, 1990.

50. Ben-Haim, Y. and Elishakoff, I.: Convex Models of Vehicle Response to Uncertain but Bounded Terrain, Proceedings of the ASME Pressure Vessels and Piping Conference (H. Chung, ed), Honolulu, Hawaii, ASME, 1989, 81-88; also, Journal of Applied Mechanics, 1990 (to appear).

51. Ben-Haim, Y. and Elishakoff, I.: Non-probabilistic Models of Uncertainty in the Non-linear Buckling of Shells With General Imperfections: Theoretical Estimates of the Knockdown Factor, Journal of Applied Mechanics, Vol. 111, 1989, pp. 403-410.

52. Leunberger, D.G.: Introduction to Linear and Nonlinear Programming, Addison-Wesley, Reading, MA, 1984.

53. Arora, J.S., Introduction to Optimum Design, McGraw Hill, New York, 1989.

54. Elishakoff, I., Gana-Shvili, Y. and Givoli, D., Convex Optimization as Applied to Uncertin Eccentricities, Sixth International Conference on Applications of Statistics and Probability in Civil Engineering, Mexico, 1991 (to appear).

55. Elishakoff, I. and Nordstrand, T., Probabilistic Analysis of Uncertain Eccentricities on a Model Structure, Sixth International Conference on Applications of Statistics and Probability in Civil Engineering, Mexico, 1991 (to appear).

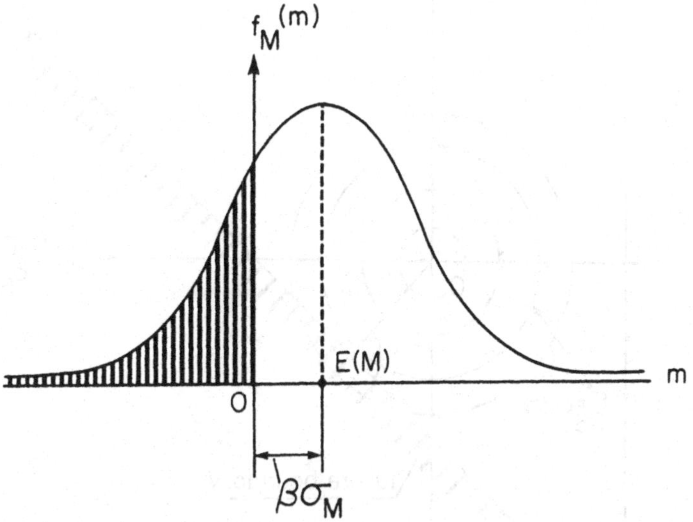

Fig. 7.1 - Probability density of the safety margin

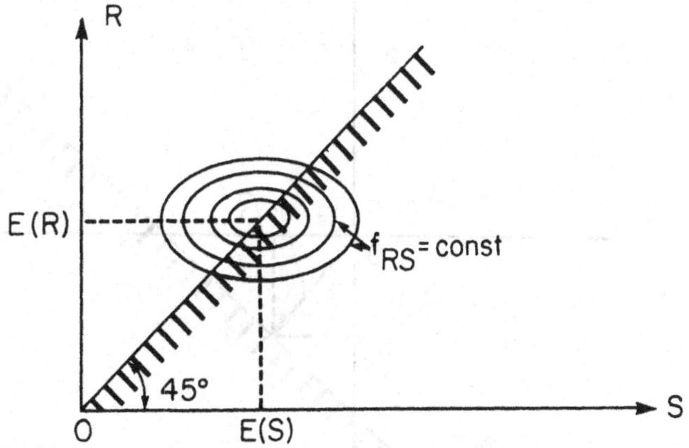

Fig. 7.2 - Curves of equal joint probability density of stress and strength

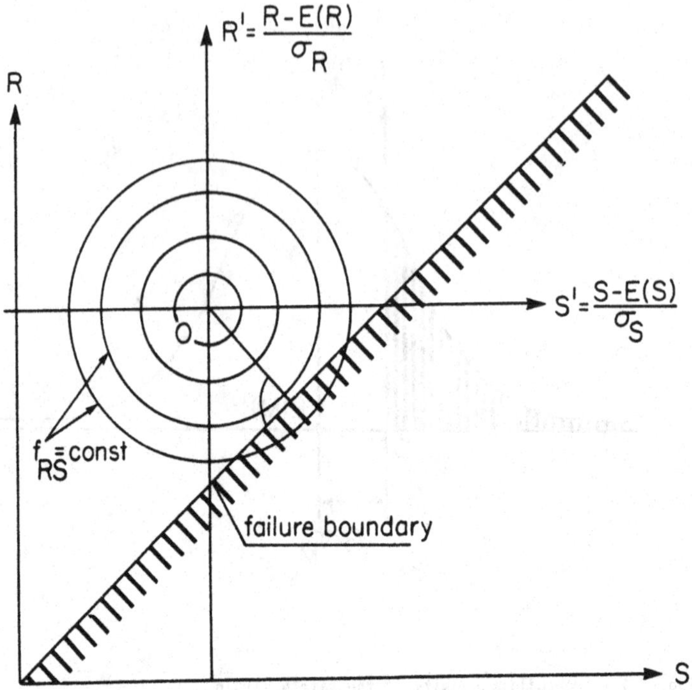

Fig. 7.3 - Curves of equal probability density in the standard space and the failure boundary

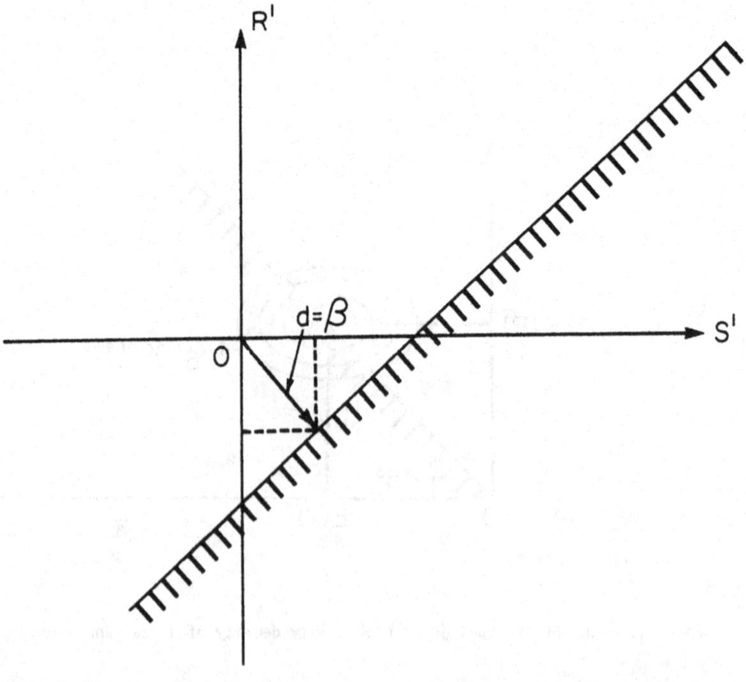

Fig. 7.4 - Interpretation of minimum distance for the linear failure boundary

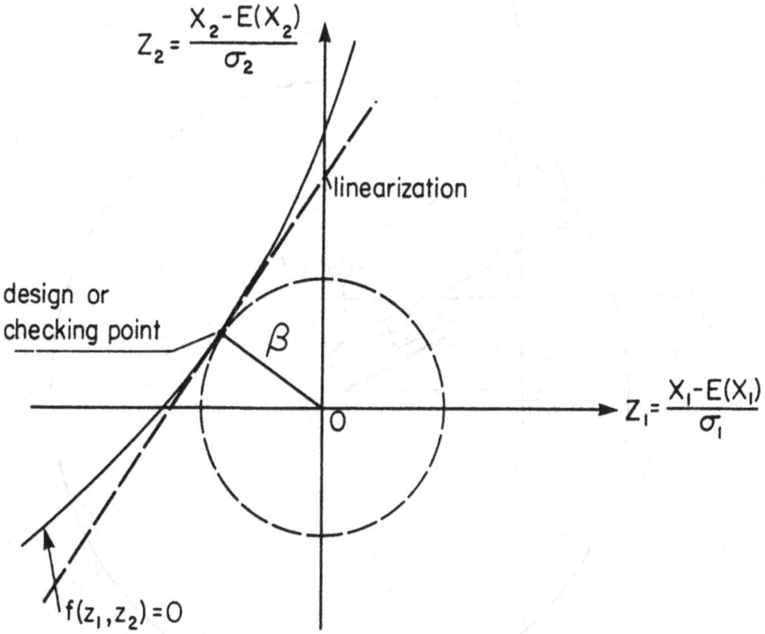

Fig. 7.5 - Minimum-distance as the reliability index

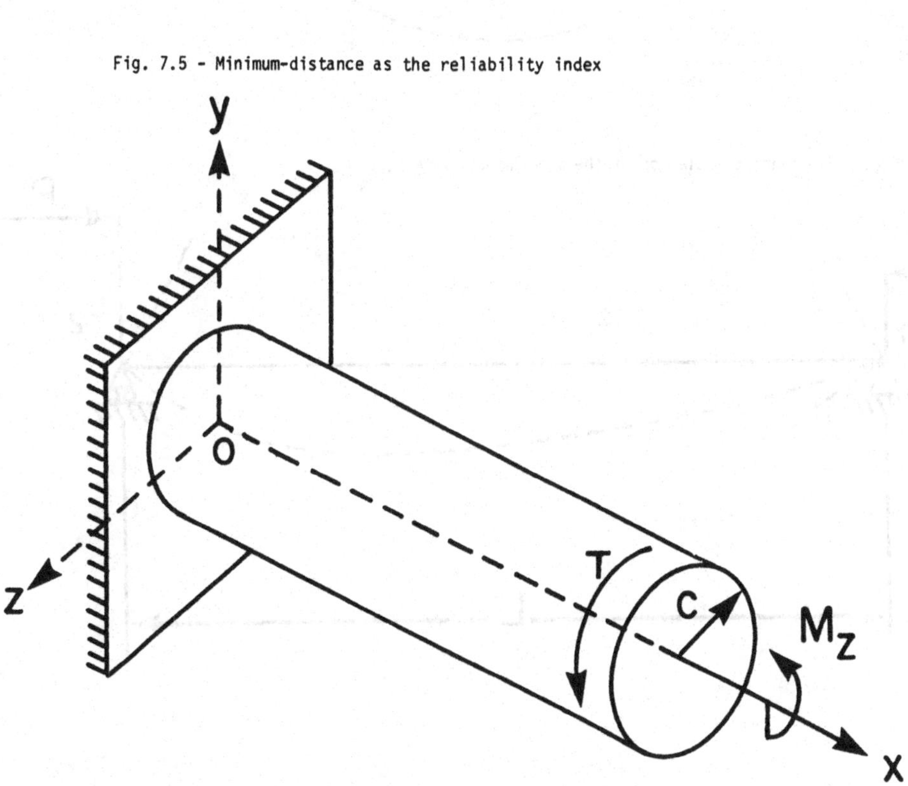

Fig. 7.6 - Circular shaft subjected to random bending moment and torque

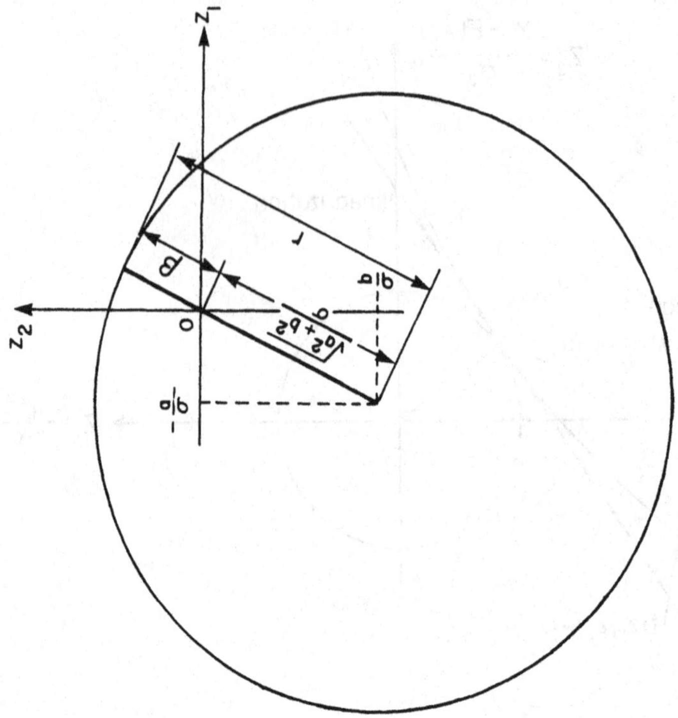

Fig. 7.7 - Exact calculation of the minimum distance

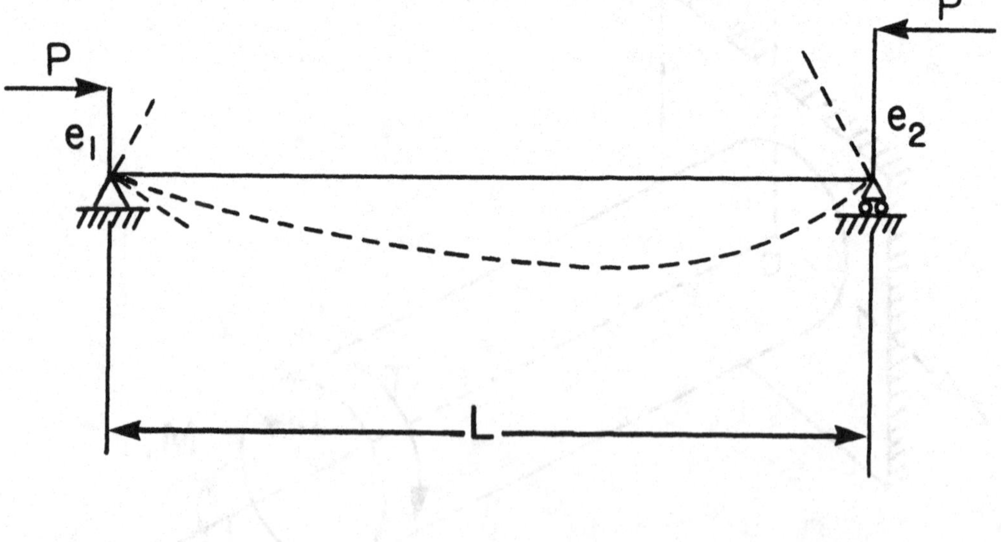

Fig. 7.8 - Beam-column subjected to eccentric forces

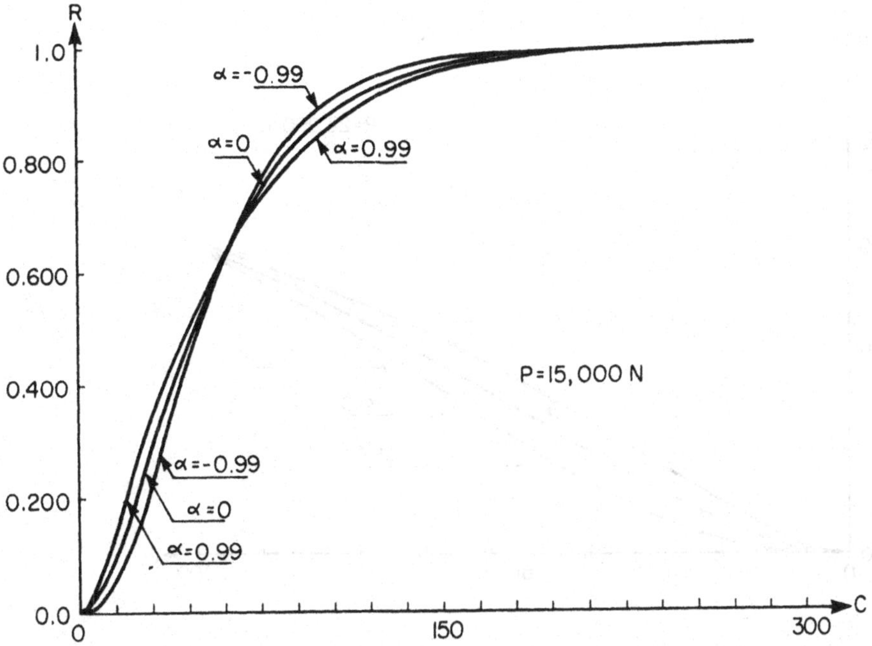

Fig. 7.9 - Structural reliability versus radius

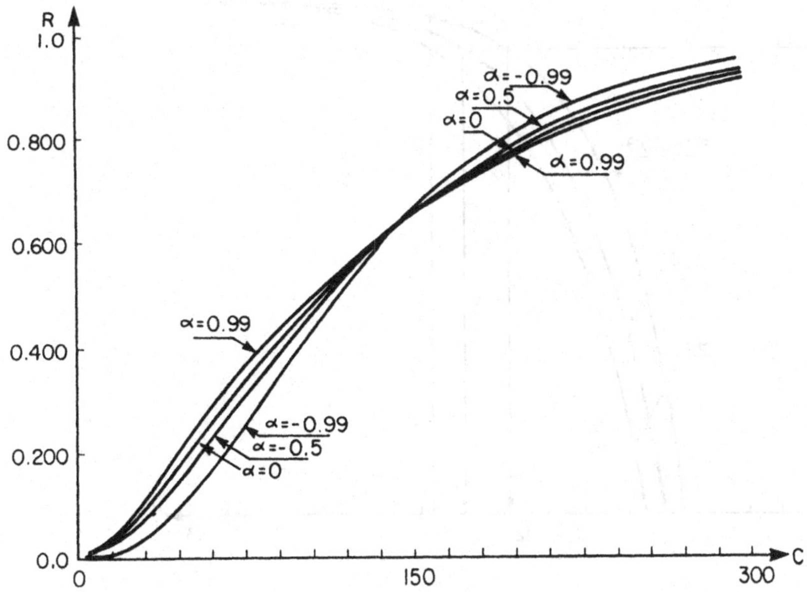

Fig. 7.10 - Dependence of reliability upon parameters

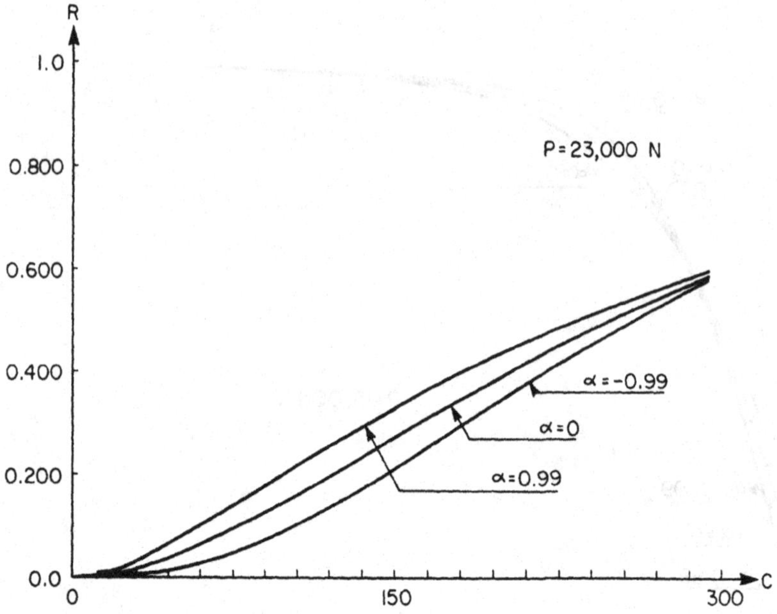

Fig. 7.11 - Influence of the correlation coefficient

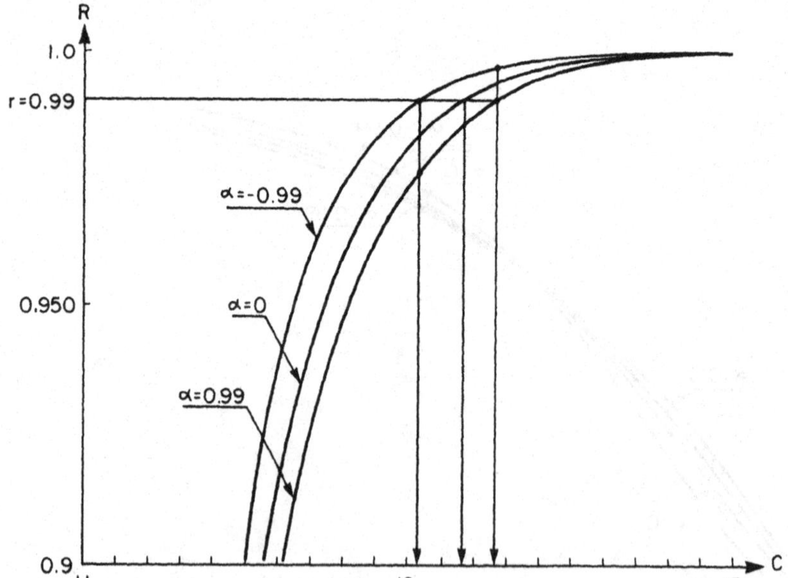

Fig. 7.12 - Solution of the design problem

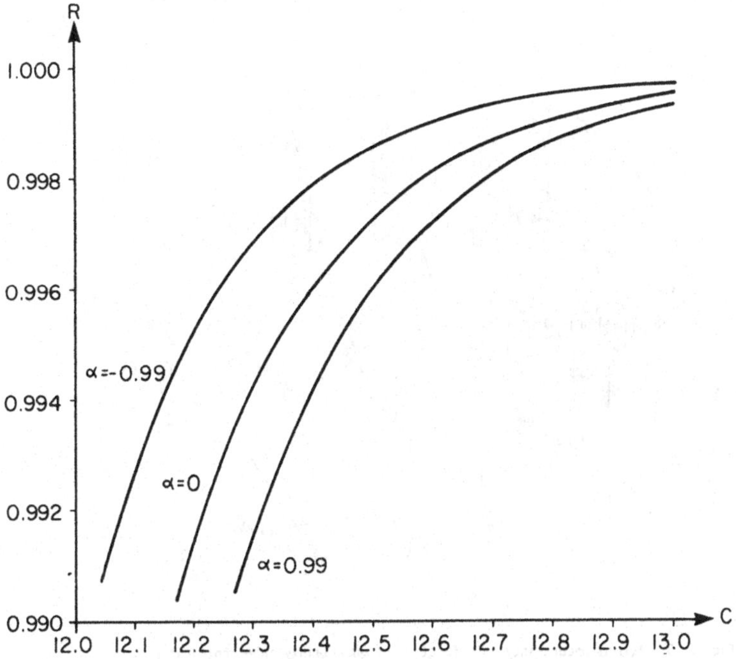

Fig. 7.13 - How does the high required reliability influence the decision-making

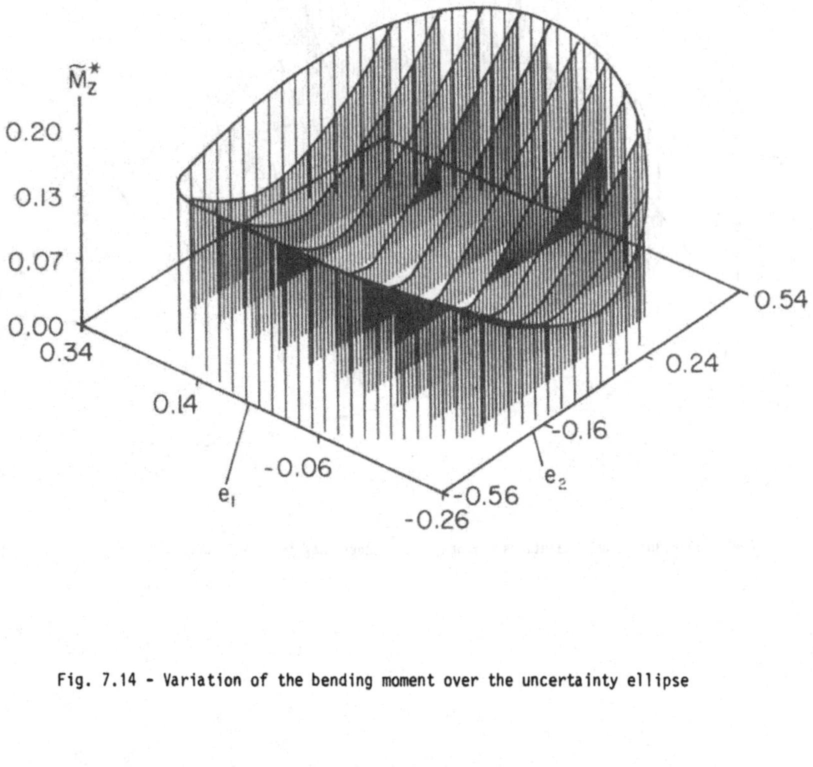

Fig. 7.14 - Variation of the bending moment over the uncertainty ellipse

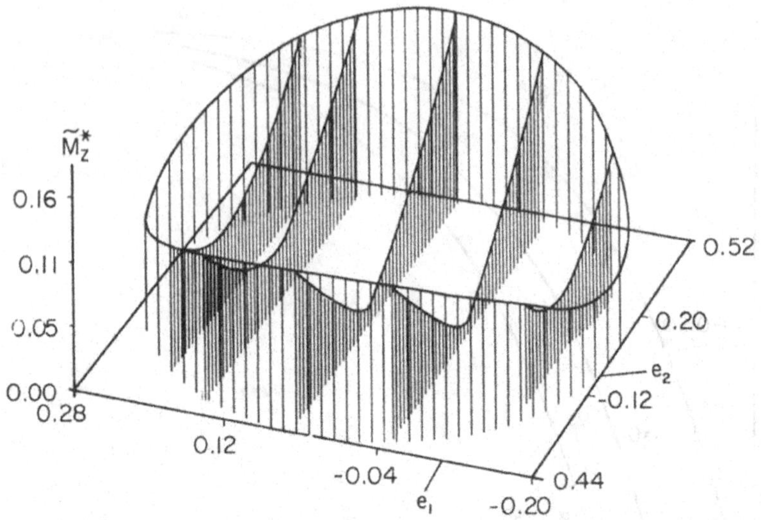

Fig. 7.15 - Moment uncertainty as a function of uncertainty in eccentricities

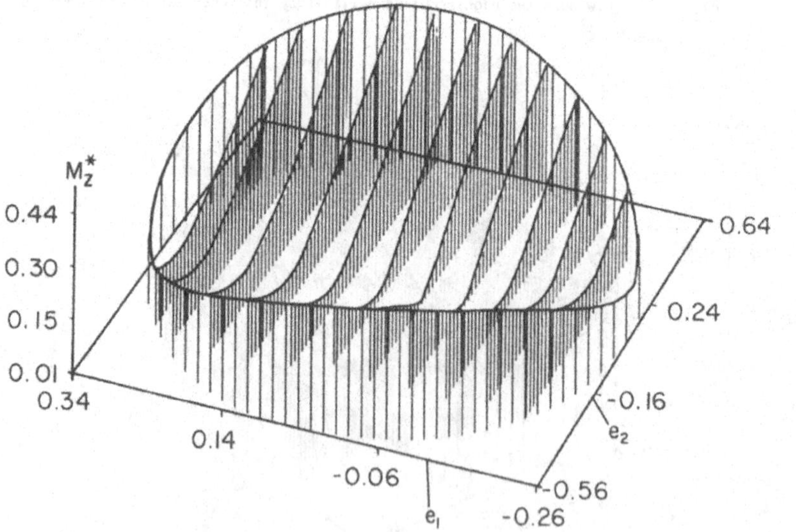

Fig. 7.16 - Moment uncertainty as a function of uncertainty in eccentricities

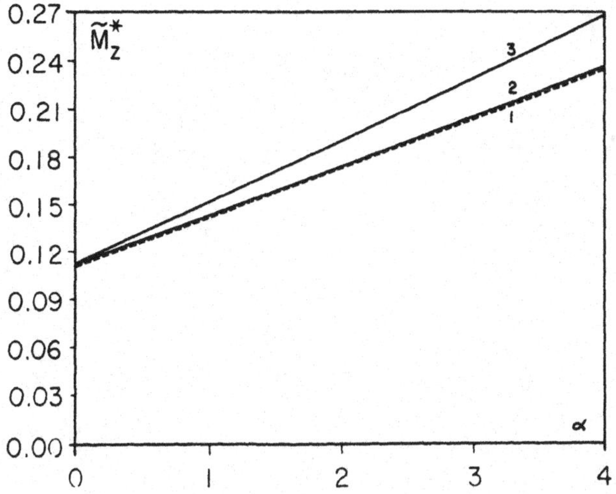

Fig. 7.17 - Comparison of first order and second-order solutions with exact solution